The Flying Adventures of
Jessie Keith "Chubbie" Miller

ALSO BY CHRYSTOPHER J. SPICER

*Clark Gable, in Pictures:
Candid Images of the Actor's Life* (McFarland, 2012)

*Clark Gable: Biography, Filmography,
Bibliography* (McFarland, 2002)

ALSO OF INTEREST

Fall Girl: My Life as a Western Stunt Double
by Martha Crawford Cantarini and
Chrystopher J. Spicer (McFarland, 2010)

The Flying Adventures of Jessie Keith "Chubbie" Miller

The Southern Hemisphere's First International Aviatrix

CHRYSTOPHER J. SPICER

McFarland & Company, Inc., Publishers
Jefferson, North Carolina

LIBRARY OF CONGRESS CATALOGUING-IN-PUBLICATION DATA

Names: Spicer, Chrystopher J., author.
Title: The flying adventures of Jessie Keith "Chubbie" Miller : the Southern hemisphere's first international aviatrix / Chrystopher J. Spicer.
Description: Jefferson, North Carolina : McFarland & Company, Inc., Publishers, 2017 | Includes bibliographical references and index.
Identifiers: LCCN 2016056343 | ISBN 9781476665313 (softcover : acid free paper) ∞
Subjects: LCSH: Miller, Jessie Keith. | Women air pilots—Australia—Biography. | Women in aeronautics—History.
Classification: LCC TL540.M525 S65 2017 | DDC 629.13092 [B] —dc23
LC record available at https://lccn.loc.gov/2016056343

BRITISH LIBRARY CATALOGUING DATA ARE AVAILABLE

ISBN (print)978-1-4766-6531-3
ISBN (ebook) 978-1-4766-2732-8

© 2017 Chrystopher J. Spicer. All rights reserved

No part of this book may be reproduced or transmitted in any form or by any means, electronic or mechanical, including photocopying or recording, or by any information storage and retrieval system, without permission in writing from the publisher.

Front cover: *inset* Jessie Miller in her flying gear after her arrival in Australia in 1928 from England (courtesy The Cobbs Auctioneers); *background* world map (© 2017 iStock) showing path of Miller's flight from England to Australia

Printed in the United States of America

McFarland & Company, Inc., Publishers
  Box 611, Jefferson, North Carolina 28640
  www.mcfarlandpub.com

For Marcella,
Who beat the odds,

For Miranda,
Who always knew how to fly,

And for Max, Slugger and Princess Buttercup
Who reached the second star on the right before us.

Contents

Acknowledgments ix
Preface 1
Prologue 5

One. Under the Southern Cross 9
Two. Into the Air 25
Three. So Close, So Far Away 45
Four. Australia Fair 66
Five. Burning Bridges 88
Six. The Powder Puff Derby 97
Seven. Flying the Ford Tour 122
Eight. Lost in the Bermuda Triangle 136
Nine. Hard Times Down South 154
Ten. An Awful Thing Has Happened 171
Eleven. England, My England 209

Epilogue: Full Circle 232
Chapter Notes 237
Bibliography 253
Index 257

"Below the earth has shed its darkness. There is the silver of countless lakes and streams. The greatest things to be seen, the ancients wrote, are sun, stars, water, and clouds. Here among them, of what is one thinking? ... of flying itself, the imperishability of it, the brilliance."
—James Salter, *Burning the Days* (1997)

Acknowledgments

This has been an ongoing project for many years, having been put on hold for some time because of the lack of interest from Australian publishers. However, thanks to the team at McFarland, Jessie Keith Miller's story can finally be heard.

I have two other people in particular to thank for the renewal of my interest in having Jessie's story published, and they are Andrew Lancaster (Bill Lancaster's great-nephew) and Noni Couell, who together persuaded me to become involved in their recent documentary film about Bill and Jessie, *The Lost Aviator*. They deserve a special word of appreciation, for without their interest and encouragement in reviving this project you would not be reading this book.

I would also like to especially thank Heather Taylor, maker of the documentary *Breaking through the Clouds* and tireless researcher of and speaker about the women, places and events of the 1929 Women's National Air Derby for her continual encouragement and invaluable assistance throughout the journey of this book that extended even as far as reviewing the appropriate chapter for me.

My grateful appreciation goes out to these people at the following organizations for their assistance with specific research and the collecting of images:

Timothy Achee, University Archivist, St Louis University Libraries
Simone Anderson, assistant to Robert Whiteman, The *Liberty* Project
Ronice Blair and Pauline Williams, Southern Cross Museum
Norm Bollen and Brian, Fort Plain, New York, historians, and Robert C. Carter, Minden Town Historian
Jenny Camilleri, Broken Hill Historical Society
Ed Coates, Ed Coates Collection and Frank Walters Collection
The Cobbs Auctioneers
Leroy Douglas, President, Long Island–Republic Airport Historical Society
Dawn Hugh, Archives Manager, History Miami

Acknowledgments

George Jenks, Manager, Avro Heritage Center
Tanya Kato, Curatorial Assistant, Special Collections, Honnold/Mudd Library, Claremont University Consortium, and Gale Burrow, Head of Outreach and Public Services, Special Collections, Claremont Colleges Library
Beverley Lee, Digital Content Management Officer, Northern Territory Government, Department of Arts and Museums
Lesley Niblett, Membership Assistant, Ninety-Nines
Howard Rub, Darling Downs Aero Club
Michael Sharaba, Collections Specialist, International Women's Air and Space Museum
Lynne Shipley, Dominic Winter Auctioneers
State Library of Victoria, Newspaper Collection
State Library of Western Australia

And I am thankful for the following individuals who offered valuable assistance:

Trish Burgess	Stuart Mortimore
Eddie Coates	John Pollock
Ashley Copeland	William John Scott
Martha Hawthorne Dixon	Crisanda Singer
Patty Hyatt	Debbie Squires
Gene Nora Jessen	Kate Tiffany
Chris Mancini	Mike Walker
Jane Marcellus	David Watson
David Marshall	Charles Woodley

Preface

Before Amelia Earhart, there was pioneer Australian aviatrix Jessie Keith Miller. Before her, no woman had ever traveled so far in the air or even crossed the equator above the planet's surface when Jessie became the first woman to fly from England to Australia, landing in Darwin on March 19, 1928, as co-pilot with William "Bill" Lancaster in an Avro Avian biplane. Earhart's transatlantic flight in June that year of some 2,250 miles as just a passenger on board the large Fokker trimotor *Friendship* pales into insignificance alongside Jessie Miller's journey of over 14,000 miles in the tiny open-cockpit *Red Rose*, over half the distance of a global circumnavigation, with only one other person. Not until May 1930 would English aviatrix Amy Johnson be the first woman to fly that route solo, and not until 1934 would another Australian aviatrix, Freda Thompson, fly it alone. Jessie was only the third Australian woman to be granted a pilot's license, the first of whom was Australian nurse Hilda McMaugh in England in November 1919, followed by Millicent Maud Bryant in Australia in March 1927. Bryant, along with Evelyn Follett, had actually attempted to join the newly formed Australian Aero Club in 1926 in order to train, but the committee didn't accept women as student pilots. Before Bryant's career was tragically cut short by her death in the *Greycliff* ferry disaster in Sydney Harbor, she took part in the first women's air race in Australia, the Ladies Oaks Race, held in October 1927, over a distance of eight miles across Sydney. The following year, the Oaks Race and the associated air pageant attracted a crowd of 100,000 people because one of the features of that day's events was the landing of Jessie Miller and Bill Lancaster in Sydney on their way south from Darwin.

    I first became interested in Jessie Miller's story many years ago when, while researching another project in the American state of Ohio, I heard a story about an Australian woman whose biplane had landed in a farmer's field outside of the small Ohio town of Xenia during a race in 1929. At that time, I could find nothing about any Australian woman flying a plane in America, which made me even more determined to discover the identity of this mystery Southern Hemisphere aviatrix. Eventually, it wasn't Jessie Miller the pilot whom I found but Mrs. Keith Miller the witness in a sensational murder trial, and that discovery led me to Ralph Barker and his book about Jessie's flying partner, *Verdict on a Lost Flyer*. Fortunately, I managed to make contact with Ralph shortly before he passed away and he was able to share with me some of his information that inspired me to search further. However, *Verdict* is focused on Lancaster and leaves Jessie essentially in the background, where she's remained ever since, yet Jessie's career as a record-breaking pioneer pilot actually rapidly eclipsed that of the careless and accident-prone Lancaster. Jessie became a significant female pioneer pilot in both American and Australian aviation history, whereas Lancaster is really only remembered today because of two major personal calamities: his trial for the shooting death of Jessie's fiancé, and his subsequent disappearance and death in the middle of the Sahara Desert.

Despite Jessie Keith Miller's significance to aviation history, the story of this remarkable Australian woman who became the first international aerial ambassador for her country remained virtually unknown in her own country until I revived it as a chapter in my book *Great Australian World Firsts*. As so often happens, though, the more I searched the more I found, far more than could be squeezed into a single book chapter. Jessie had always been interested in writing and had always wanted to tell her own story, yet publishers have been only interested in the more sensational events surrounding Lancaster. So, although she made at least two attempts at books about her flying career, no major published work is known to have eventuated, and despite frequent mentions of diaries, journals and log books having been written by both Miller and Lancaster, the original volumes have never been located. Only Jessie's notes, interview transcripts and articles have survived and the versions of Lancaster's diaries published in newspapers. Educated in the arts, Jessie Miller began traveling and writing while married to an experienced journalist, and she continued to write and be published in American, Australian and English newspapers and magazines throughout her flying career, so even though a book on her life has not been published until now, she did leave behind quite an extensive autobiographical record to which I refer when using the terms "recollected" or "remembered" or from which I have drawn dialogue or mention her thoughts. Along with the interview that Ralph Barker conducted with her only a few years before she died, these long-forgotten accounts are now the only chance to hear Jessie's voice telling her story, and so I've relied on them extensively here as primary sources to give Jessie a chance to express her opinion and point of view. Like many women pilots of her era, Jessie found that much of her life became public news and, in fact, to a certain extent that was quite deliberate. Flying cost money, and to raise that money an aviator had to draw attention to their flying in order to attract sponsors and funding. Even Amelia Earhart, married to the wealthy George Putnam, still had to sign deals with Lucky Strike cigarettes and market her own line of clothing and luggage to fund her career. So, much of Jessie's flying career was covered extensively by newspapers in the U.S., Australia and England, although not always accurately, and so given the lack of letters and diaries I have used these accounts as well so that Jessie's story can be understood in the context of her time and so that as much of Jessie's story as possible can be told from her point of view, that of an independent career woman who chose her own path through the world.

For Jessie Miller, one record was never enough, and she continued to break records throughout her flying career. She was the first woman across the Mediterranean in the air, the first woman to travel by air through the Near East and Middle East to the Far East, then through what is now Indonesia to be the first woman across the Timor Sea by air. She was then the first woman by air across the Australian continent from north to south and the first woman to pilot an airplane across Bass Strait to Tasmania. She was the first female pilot to fly solo from the east coast of the U.S. to Cuba, the first female pilot from the Southern Hemisphere to fly competitively in the U.S. and to compete in national air races there, the first woman to be granted a commercial pilot's license in Canada, and in 1930 she became the second official female holder of the American transcontinental flight speed record in both directions and the first woman to hold that record as a result of unaccompanied solo flight. A personal friend of other pioneer female pilots such as Amelia Earhart, Amy Johnson and Nancy Bird Walton, of aircraft designers and engine builders alike, Jessie Keith Miller was known and respected for many years as Australia and the Southern Hemisphere's first international aviatrix.

One of the probable reasons her memory has faded from Australian history is that Jessie last visited Australia in 1928 and then never returned, although she certainly intended to. Most of her association with aviation was in America, where she is still honored as one of the charter members of the first female aviators' organization, the Ninety-Nines, and as one of the select group of female aviation pioneers who competed in the first American national air race for women, the 1929 Women's National Air Derby, or the "Powder Puff Derby," as it became colloquially known.

However, one of the other possible reasons for Jessie's omission from Australia's early aviation history records was her involvement in one of the great sensational murder trials of the 1930s, during which Jessie and her married pilot colleague, Bill Lancaster, had to admit they had an affair during their England to Australia flight. Nevertheless, despite Bill continuing to profess his love for Jessie, he would not divorce his wife and so Jessie eventually fell in love with her younger, dark and handsome biographer, Haden Clarke.

Unfortunately, as was revealed later, he couldn't sort out his romantic affairs, either, but it seemed at the time that he was able to marry her. Consequently, the prosecution claimed, a jealous Bill murdered his rival. Or consequently, the defense claimed, a depressed Haden Clarke took his own life because he'd lost hope. Jessie had to choose a side. Although Jessie was undoubtedly an independent woman who made her own choices about her life, both American and Australian society of that period still clung to conservative moral codes, and it was obvious that Jessie was invited by the United States Department of Labor to leave America after the trial for infringing those codes. A young country such as Australia, self-conscious and sensitive of its international public image, would have been embarrassed to be associated with a woman mixed up in such a public moral scandal, so Jessie was quietly airbrushed from early Australian aviation history.

Almost certainly one other reason Jessie faded from memory was that her aviation career did not survive World War II, although she was an early advocate of defense forces employing women as transport pilots. By the time the war was over, Jessie had married another well-known pilot and retired from public life. She was, after all, a woman from a conservative family background who had endured a traumatic, life-changing experience during which her inner and intimate self had been exposed to public gaze for question and comment. Now, growing older and having come close to losing her life on a number of occasions, Jessie took the opportunity to trade adventure for a quieter and probably longer life. Only the unexpected discoveries concerning Lancaster not long before her death would bring Jessie momentarily out of that retirement.

Yet in only a few years, pioneer aviatrix Jessie Keith Miller made a major contribution to Australian and American aviation history, for which she is still honored as a founding member of the Ninety-Nines and in the Women in Aviation Pioneer Hall of Fame. She still awaits recognition in her own country. In blazing the trail for women aviators from England to Australia, she became an early symbol of what women could accomplish. Many future women pilots read about her adventures in newspapers and followed her journeys on maps pinned to their walls, saying to themselves, "If Jessie can do that, then so can I."

After so long, this is her story. Jessie Keith Miller was a pilot once...

Prologue

In the early coolness of that green tropical morning, Jessie Miller stood with her hands on her hips in the tall grass of a field on the island of Bangka in the Dutch East Indies, now Indonesia, quietly and intently studying the ridiculously small biplane in which she and her pilot friend Bill Lancaster had flown here on such fragile wings. It was already so humid that you could practically see droplets of moisture forming in the air, but Jessie was doing what she had done every morning for the past three months since she and Bill had left England to fly to Australia: carefully inspecting their plane. As they traveled south through these islands on their way to Australia, they would be flying over places few, if any, Europeans knew existed outside of adventure stories. The island on which she stood is reputedly one of the settings for Joseph Conrad's novel *Lord Jim*, for example.[1] Even in this Year of Our Lord, 1928, it would be a rare person out here in this jungle island outpost of the Dutch East Indies who'd even seen an airplane, let alone been in one. The only winged objects usually inhabiting these skies had beating hearts and were covered in feathers rather than fabric.

With a wingspan of 28 feet and about 24 feet long, Jessie and Bill's Avro Avian Mark III biplane, *Red Rose*, was probably about the same size from wingtip to wingtip as the average small, modern hang glider; many of the larger hang gliders, in fact, are far bigger. The Avian was supposed to have an average ceiling of about 18,000 feet, but biplanes of that era are just as susceptible to the vagaries of updrafts and bad weather as hang gliders and large birds and so are usually flying much lower. This particular flying creature's skeleton was made of wood covered with a "skin" of tough fabric coated in a glue-based waterproof mixture known as "dope," and its tiny heart was the 85hp four-cylinder engine that drove it through the air at a top speed of around 100 miles per hour, if it was working perfectly and there was a tailwind. A really strong headwind could literally push them backward, so the average cruise speed was about the same as the horsepower: 85 miles per hour. The entire plane weighed only 1700–2000 lb fully loaded, which gave a range of about 400 miles with the standard 35-gallon fuel tank, and it was piloted using just a few basic instruments that did not include a radio. If you wanted to know where you were, you consulted your flight map, leaned over the edge of the cockpit and looked down, or simply landed and asked someone. To communicate with the person in the other cockpit, you wrote a note and passed it to them or shouted through a speaking tube. To communicate to another plane, if you ever saw one, you flew close to them and used sign language.

Jessie began the habitual slow walk around the aircraft that she had carried out every morning of their journey, never taking her eyes off it, checking carefully for any minute tears or wear spots in the fabric that would be likely to rip open in the air, making sure the bracing wires were taut and without any signs of fraying and that wing struts were rigid and that clips and nuts were firmly tight. In machines as small as these, you shared an intimacy with your environment that came at a price: one small mistake—one wire a little slack or a nut worked loose—could kill you. There was no room for a parachute in the tiny cockpits; you were often flying too low for one to be really effective anyway. Having completed her inspection without finding anything amiss, she would have stepped back and relaxed a little, reassured for the moment. She and Bill had landed on this grassy field the previous evening, having flown south from Singapore where they had spent the weekend. It had been raining so hard when they landed there on Saturday that the engine had flooded and stalled, fortunately after they had touched down. Then so many people converged on the plane that the field was quickly reduced to a muddy quagmire in which the plane's wheels became bogged; unable to taxi, she and Bill had to climb out. By then, though, they had firmly established the world record for distance flown in a light aircraft. No one else had flown so far in a plane that small, certainly no woman.

The *Red Rose* had been the first small airplane to reach Singapore from England, making Jessie the first woman to fly from there to the Far East. As her regular reports had been telegraphed back home to newspapers as they had steadily flown southeast, Jessie had become a beacon of hope for young women hoping to fly a plane someday. As the police held people back from the plane, Jessie and Bill had climbed out to be welcomed by an official party that included the Colonial Secretary, Sir Hayes Marriott, who was Acting Governor at the time; Major Lowe, representing the Commanding General; and Mr. R. Johnson, who was the captain of the Singapore Flying Club. As they stood there in the rain, shoes filling with water, an irate Bill Lancaster had left them in no doubt about what a dreadful and dangerous landing field this was for a colonial center such as Singapore. In fact, he later told a *Straits Times* journalist, it was one of the worst landing grounds they had yet encountered on the trip. Naturally worried that they might not be able to take off again if the weather didn't ease, Jessie and Bill had discussed with officials whether they should move the plane to the nearby Serangoon Road racecourse, but the Turf Club insisted that area was just as prone to flooding, and so they had tied the *Red Rose* down for the night and hoped for the best.[2]

Dressed in her tropical flying outfit of khaki shirt, shorts and topee (pith sun helmet), Jessie went through the weary motions of answering inane questions from journalists who persisted in seeing a woman rather than an aviator. However, knowing the value of public relations, she gritted her teeth and once again recited that yes, she was looking forward to returning to her home in Melbourne and that, yes, she did regularly fly the plane solo and that she was indeed quite capable of refueling the plane herself and of dealing with some of the mechanical problems. In fact, knowing they were about to fly a long leg out over the open water of the Timor Sea to Darwin, they would be carrying out engine maintenance most of the next day. So, after a luncheon hosted by the Flying Club at Raffles Hotel and dinner that night with Sir Hayes and Lady Marriott, followed by a good night's rest, Jessie and Bill had spent Sunday overhauling the engine and closely inspecting the *Red Rose*. From here on, it would be hard flying, and they wanted to make sure everything was going to work just the way it should.

By Monday morning, 9 January 1928, the weather had lifted and so Jessie and Bill took off from the airfield and circled Singapore Harbor before turning the nose of the plane south towards the port of Muntok on Bangka Island. As they followed the east coast of Sumatra south, it occurred to Jessie that she was now the first woman to cross the equator in an airplane. Looking down, she thought that the dense jungle below her seemed to merge into one continuous, seamless, wide green rug that rose and fell under her, leaving no impression of depth. Palm-fringed bays alternated with mangrove swamps, and tiny emerald islands with lacy white edges dotted the coastline. Except for a herd of elephants she spied through a narrow break in the trees, there were no other living creatures to be seen. Eventually, she and Bill left the coast behind as they flew briefly out across the South China Sea before crossing the northeast tip of Bangka Island to approach the port of Muntok, or Mentak as it is also known, founded in 1732 as a center of tin mining and provincial administration. As they flew over the open, sloping strip of land where they planned to land, Jessie and Bill could see a crowd of Dutch settlers who had turned out in force to meet them, but they could also see that the grass had not been cut and was at least thigh high, making it impossible to see any dips or hollows. With little choice, they just had to cross their fingers and put the plane down. Fortunately, they landed safely to a warm welcome from the local Dutch people, and that night some one hundred people attended a dinner in Jessie and Bill's honor. However, although a new radio set had been shipped in for the occasion, it could not be persuaded to work and so they had not been able to contact anyone to communicate their whereabouts.

So, that morning Jessie was considering the implications of their communication failure as she was inspecting the plane when Bill finally showed up, looking his usual jaunty self as though he didn't have a care in the world. They fueled up the plane and loaded their things on board. Bill climbed in, Jessie spun the prop until the engine caught, and then she clambered into her usual place in the forward cockpit. While everybody waved and cheered, Bill taxied the *Red Rose* into the wind and they took off through the grass, lifting smoothly into the cool morning air. As the plane climbed to about 130 feet, Jessie was just leaning over to wave back to the crowd when, without so much as a warning cough, the engine abruptly cut out. There was only the sound of the wind rushing by. For a few seconds, Jessie thought Bill had left something behind again and was turning back to land, until she realized he was trying to start the engine, again and again. There was nowhere for them to go. Ahead was the sea. Behind them were hills and houses. With a sickening sideways lurch, the *Red Rose* fell out of the sky, fluttering towards the ground like a leaf from a tall tree. With no engine power, Bill had no flight control. The only thing he could do in an attempt to save their lives, especially Jessie seated up front under the wing fuel tank, was to keep one wing down so it would act like a shock-absorbing crumple zone as they sideslipped into the ground, rather than go in nose first and have her killed by the engine. As they fell, time slowed for Jessie; it felt as if hours crawled by until finally the ground came at her in a rush and they smashed into it with a grinding crash, somersaulting down the slope. The undercarriage ripped off as the plane flipped over and then Jessie's world went very still.

After a few seconds, she shook her head to clear it from the shock of the impact and discovered she was hanging upside down, imprisoned in a web of tangled wires and wreckage and being drenched in petrol from the tank that was now above her. Hearing the petrol sizzling as it hit the hot engine, Jessie imagined an imminent flaming explosion that would incinerate her "like a witch at a stake" and began a desperate struggle to extricate herself. When she found enough breath in between twisting, tearing and pulling, she cried out to Bill, but there was no response.[3]

ONE

Under the Southern Cross

The wide, flat, dusty streets of an outback gold-mining town bleeding into the vast red emptiness of inland Western Australia are a long way from the dense green growth of tropical island jungles.[1] In fact, the small town of Southern Cross where Jessie Miller was born, some 228 miles northeast of Perth, the capital city, is a long way from anywhere. It was, however, the first place where gold was discovered in what became known as the Eastern Goldfields of Western Australia, and so Southern Cross predates the bigger and more well-known mining towns of Coolgardie and Kalgoorlie farther to the east. In 1888 three prospectors, Thomas Risely, Mick Toomey and Charlie Crossland, used the major Southern Cross constellation, or Crux, to navigate across the plains towards a place called Kookoordine in search of water.[2] They didn't find their water then, but they returned later after the rains came, directed by the same constellation, and this time they found a reef of gold as well, so in honor of the stars that brought them such good fortune they named the area Southern Cross.

For a few short years, the town boomed with a gold rush, but then in 1892 Arthur Bayley rode into town with a saddlebag containing 554 ounces of gold that he'd discovered at Fly Flat, now known as Coolgardie. As the news spread through town like a brush fire, every able-bodied man left for the Flat, beginning the biggest stampede for gold in Western Australian history. So, instead of developing into a mining center, Southern Cross became the center of administration for the local municipality, Yilgarn, and a courthouse was built and a road constructed to Coolgardie for the stagecoach. Then when the railroad finally arrived in 1894 Southern Cross was firmly linked to the coast in one direction and to the wealth of the goldfields in the other.

A town looking for a future like Southern Cross needed a bank to assure and insure its financial resources and that's why Jessie's father, Charles Stanley Beveridge, came to the town in January of 1898.[3] When Southern Cross was first established as a mining camp, there wasn't enough hard currency in cir-

The remote Western Australia town of Southern Cross, where Jessie was born, photographed by Ernest Lund Mitchell, c. 1910 (image No. 008342PD sourced from the collections of the State Library of Western Australia and reproduced with the permission of the Library Board of Western Australia and the State Library of Western Australia).

culation in the area to maintain a daily supply of it and so the mines paid their workers with paper IOUs that the stores accepted, writing their own IOUs as change. However, as the town grew, this honor system quickly became too unwieldy and so in 1890 the Commercial Bank established a branch in the town, not only to manage currency but also to provide a safe repository for the miners' gold until it could be sent to Perth. The first manager, Mr. Charles Saw, carried out business from a small two-room iron and canvas building that once caught fire and another time was un-roofed by a "willy-willy," or small tornado. Nevertheless, the bank endured and by 1892 deposits there had reached £7,000, while the bank purchased some £50,000 of gold a year from miners. Obviously a more substantial bank building was required, and so a two-story structure was built using adobe bricks and stone, with the bank manager's residence on the upper floor, and it was into that residence Charles Beveridge moved as the new manager of the bank literally under his feet.[4]

    Charles came from an English family who could trace their origins back to the thirteenth century and who, during the seventeenth century, had settled mainly around the neighboring Medway River ports of Chatham and Gillingham, Sir Francis Drake's hometown. Charles' family, however, had then moved west to Portland, Dorset, where he was born in 1866. Although his grandfather was a bishop and his father a clergyman, the young Charles found a job with the London and Country Bank. Perhaps he felt a need to do something different with his own life because most of his family had lost their lives while working for the church. His father, the Rev. Thomas George Beveridge, had been a minister for the Congregational Church when in 1872 he journeyed with his family to Madagascar as a missionary for the London Missionary Society. After five years there, Thomas' health deteriorated and so the family was returning to England on the SS *Cashmere* when it was wrecked near Cape Guardafui on the Horn of Africa.[5] Charles' father, mother Emma, brother Herbert and two sisters Jessie and Beatrice all drowned; only his younger brother, Sidney, survived the tragedy itself, along with Charles and their sister Nellie, who were in boarding school in England. Unlike his brother Charles, the Rev. Sidney Alexander Beveridge followed family tradition and became a missionary for the London Missionary Society in Samoa, where, during a seven-year stay, he developed a scholarly interest in the Samoan language. After returning to Australia, he ministered in Horsham before becoming Chaplain for the 5th Division AIF during World War I. He later returned to England and became Rector of Pinxton, Nottingham.[6] Charles' sister Nellie married Captain Thomas Hunter in 1893, only for him to be lost at sea the following year. She moved to South Africa, where she married a minister's son, John Valentine Oates, and their son Thomas Herbert Beveridge Oates became a polar explorer aboard *Discovery II* before joining the Royal Navy in England. He would be lost at sea in 1941 during World War II.[7]

When Charles decided to wed, he, too, couldn't escape the family vicarage tradition, marrying a clergyman's daughter, Ethelwyn Maude Lavinia Hall, in September 1900, at Christ Church in Southern Cross, for which occasion the bride was dressed in white China silk trimmed with lace. Maude, as she preferred to be called, was the youngest daughter of the Rev. J.K. Hall, the Church of England vicar at Wangaratta in Victoria at the time.[8] Two years before, her older sister, Jean, had married Victor Black, Warden and Magistrate for the local Southern Cross area, and her brother John lived in Coolgardie, so its' highly likely that Charles met Maude while she was visiting her relatives. Their first child, Jessie Maude Beveridge, named after her maternal grandmother, came into the world a year later to the month on Friday the thirteenth in her mother's bed above the bank. A little over a year later, the family welcomed her sister, Eleanor Jean, but it was to be only for a short while; she died in November 1903 and was buried in the local cemetery. There would have been little anyone could have done for a sick child in those days; they were far from medical assistance other than a local doctor.

Immersed in a liquid pool of heat under an endless azure sky, the town's corrugated iron and canvas buildings lay under a pall of dust stirred up from the wide, unsealed streets by people, wagons and horses and from the relentless pounding of mine batteries, the resounding thump of which provided a constant rhythmic background to life in Southern Cross day after day.

This tiny outpost of European life was situated at the very edge of the known Western Australian world; Kalgoorlie lay to the east, but to the north there was nothing but a vast expanse of barren wilderness and desert populated by indigenous native peoples of whom little, if anything, was then known. Western Australia is the second-largest subdivision within a country in the world, the equivalent in area of the country of Kazakhstan, for example, or of everything west of the American Rocky Mountains.[9] Even today, within its 1,021,478 square miles there are only some 2.6 million people and those mostly in the southwest corner, a long way from Southern Cross, where the landscape remains much as it was when Jessie lived there. From the windows of her sheltered and religiously confined home, from where she was interminably driven off to church, Jessie would have been able to see the very curve of the horizon of her world, a horizon few white people had traveled beyond and from where fewer had returned to tell the tale. But that kind of horizon can have a mysterious attraction, and perhaps even then Jessie longed to fly beyond it. A child born with such a heritage of journey in the world was always going to see beyond church walls and the boundaries of small towns.

Her chance came sooner than expected. In 1905, the Commercial Bank decided to close their branch in Southern Cross and so the Beveridge family

The grave of Jessie's sister, Eleanor, in the Southern Cross cemetery (courtesy Pauline Williams).

moved to Claremont in Perth, the capital city of Western Australia, where a year later Ethelwyn gave birth to a boy, Thomas, with whom his sister formed a close bond as the two children grew up together. Two years later, the bank transferred Jessie's father and the family moved to a larger mining town, Broken Hill, halfway across the country to the east in a different state, where this time they moved into their own house on Thomas Street. During the next nine years Jessie would grow up there, educated at the Convent High School along with her brother, Tom, where it soon became apparent by her high marks in singing, piano and theory that she could have a music career ahead of her.[10]

But in 1916 the bank suddenly moved the family to New Zealand in order for Charles to be manager of their new branch in Timaru, a major agricultural commercial center on the east coast of the South Island and the second-largest fishing port in New Zealand. Jessie attended the exclusive Craighead School there that had been established only five years previously by the Shand sisters to produce refined, cultured and capable young women, where she was a popular athlete, tennis player and prizewinning sportsperson.[11] They were only in Timaru for three years, however, returning in 1920 to Melbourne, where her father was promoted to work in the bank's head office. For Jessie, though, it was one move too many. Having had to leave teenage friends and familiar landscape behind when the family left Broken Hill and then do it once again just when she'd begun to carve a niche for herself at Craighead, now Jessie was alone in a big city, unsettled and restless, tired of the constant moving between houses and schools and weary of her family's conservative lifestyle. The constant Bible reading and churchgoing was too confining and repressive for an independent young woman with a mind of her own and so, when Jessie met the tall, rather serious *Weekly Times* newspaper journalist Keith Miller, she seized her chance to leave the nest and, without thinking twice, said "yes" when he proposed. They married quietly at Christ Church, in the Melbourne suburb of St. Kilda, on 3 December 1919. Jessie had only just turned eighteen and her husband was twenty-three. It didn't work out.

"We were quite maladjusted," Jessie would remember many years later. "It was like two babies getting married. Our characters were poles apart." Keith certainly seems to have loved her in his own way, but the young Jessie tended to the pragmatic when it came to romance. After the first few years, they were just idling along, making the best of things, sitting it out day by day as good friends who were no longer in love. Neither of them at any time blamed the other. Jessie later readily agreed that she was temperamental while Keith was calm and that, while he wanted a wife at home although he didn't like children, she had wanted desperately to go beyond her horizon and see the world and have "the right to live my own life." As time went on, it became more obvious that they just weren't suited to each other. "Let us say," Jessie said, "I had fallen out of love with him."[12]

Then, without warning, a series of tragedies struck the Beveridge family. Charles was diagnosed with throat cancer, and with little effective medical treatment available for that in those days he passed away in 1925, aged only fifty-seven. Barely two years later, her beloved brother and closest friend, Tommy, who had trained at Jervis Bay Naval Academy and become a midshipman in 1922, fell ill with cerebral meningitis and died aged only twenty-one. Jessie was inconsolable. Until the day she died, she always kept with her a photograph of Tommy dressed in his hat and uniform with its high stiff collar. He'd been a kindred spirit to Jessie; they were both adventuresome and often used to lie on the hearth rug in front of the fire and plan how they would work their way around the world together. Even though he was younger, Tommy had given her life direction and now, with both him and her father gone and her marriage drifting aimlessly from day to day, there just didn't seem much point to life. Saddened and feeling alone, Jessie felt absolutely wretched. She couldn't go on like this, she decided; she had to have something to live for.[13] At this point in her life, as Mike Walker so aptly put it, "she didn't know what she wanted, but she did know exactly what she didn't want, which was everything she had."[14]

Finally, she had an idea. Her father had always intended to return to England one day to see his homeland and family. His sister had often suggested they visit her; now Jessie could go in her father's place and perhaps that would be the solution to her depression. Elated, she persuaded Keith that it was absolutely necessary to her well-being that she go on this trip and even offered to help raise her own funds for it. She tried to persuade him to come with her to share her experience, but for reasons of his own he declined.[15] Nevertheless, always a woman who acted on her ideas, Jessie took a job selling carpet sweepers door-to-door and quickly saved the money for her ticket. Australian journalist Donald Mackenzie wrote in 1931 that four years previously Jessie, wearing an infectious smile, had "knocked at the entrance to my [Melbourne] flat, thrust a neat, suede-shod foot between the door and the sash, and refused to remove it until I had agreed to buy a new-fangled carpet sweeper that I did not want." Seeing her at a club later, he asked a female friend who she was and claimed she replied, with rather suspicious divination, that "she is a hurricane saleswoman.... And ambitious!!! ... One of these days she's going to grab the world by the whiskers and shout in his ear: 'Look at me! I did it! I'm 'Chubby' Miller!'" Years later, he still remembered her as the airwoman with the carpet-sweeper formula for success.[16]

So, in early 1927 Jessie and her friend Margaret Starr, with whom she had already traveled around the Pacific islands, boarded ship for the other side of the world with third-class tickets grasped in their hands, having planned for a six-month stay in England.[17] Arriving in London, they found a flat to share on Baker Street and were soon making friends among the local Australian expat community, one of whom was dentist Mervyn Ryland. One evening, Ryland asked Jessie if she'd like to come to a party. A friend of his would be there, he explained, a rather dashing World War I fighter pilot by the name of Bill "Bronco" Lancaster who had once lived in Australia and was now planning to be the first person to fly there in a light plane. Ryland thought she and Lancaster would have lots in common to talk about, and indeed Jessie didn't need to be asked twice. Once introduced to Bill, she quickly hit it off with the tall, few years older, slightly balding, rather careworn aviator with an attractively infectious laugh who was full of plans for a flight to Australia and interested in what she thought about them. However, it was hopeless trying to hold a conversation over the noise of the party and so they agreed to meet the next day at four o'clock for tea at the Authors' Club in Whitehall.[18]

Jessie always said Bill began planning his flight to Australia during World War I, but it's likely he was prompted into action post-war, like many pilots at that time, by those two well-known motivators: fame and fortune. While there were few surviving long-distance fliers, the ones who had survived were treated as heroes and for them Australia remained a pinnacle of achievement. Apart from a global circumnavigation, the longest continuous airplane flight is still the approximately 11,000-mile (18,000 km) journey from England to Australia. When in 1919 the Australian government had offered a £10,000 prize to the first Australian airmen who could fly an aircraft built in the British Empire from England to Australia, only two aircraft out of six entries completed the distance. Captain Ross Macpherson Smith, World War I air ace and pilot for T.E. Lawrence ("Lawrence of Arabia"), and his brother Lieutenant Keith Macpherson Smith along with their mechanics, Sergeants Wally Shiers and Jim Bennett, took off from Hounslow Heath on 12 November 1919, in a converted twin-engine Vickers Vimy bomber to fly the distance in less than 28 days, an actual flying time of 135 hours 55 minutes, for which the Smith brothers both received knighthoods.[19] They were followed by Lieutenants Ray Parer and John McIntosh, who left England on 8 January 1920, in a single-engine, two-seat Airco DH 9 bomber, after the event had actually been won, arriving as the only other aircraft to successfully complete the race on 2 August 1920. They reached Australia 206 days later where they each received a consolation prize of £500 and were awarded the Air Force Cross.[20] In 1926, Sir Alan Cobham and his crew in a De Havilland DH50 had made the first England to Australia return flight, during which his engineer, Arthur Elliot, was shot dead near Baghdad.[21] Then in May 1927, Charles Lindbergh, the "Lone Eagle," emerged from virtual obscurity to international celebrity by flying solo non-stop in his Ryan monoplane *Spirit of St. Louis* from Roosevelt Field, Long Island, 3,600 miles (5800 km) across the Atlantic to Le Bourget Field, near Paris, in a record-breaking 33 hours 30 minutes. It was the era of adventure in the air, and Bill wanted his share of it while the opportunities were there.

So, Jessie and Bill met again the next day at the club and they settled in amid the cozy wood-paneled ambience to talk. Bill had only been out of the Royal Air Force for twelve months after five years on the active list. He'd learned to fly in 1917 and had been in combat during World War I. Bill was much closer to Jessie in age than her husband, having been born in February 1898. His father, Edward, was a well-known civil engineer who had three children to a first marriage and then fathered three more, including Bill as the eldest son, when he married again to his first cousin, who by a remarkable coincidence was also named Maud. Now, by any standards Maud had a reputation for being somewhat eccentric, but nevertheless, she was artistic, fluent in several languages, the author of a popular home handy-woman book, *Electricity in the Home*, and she'd been prepared to become a mother a second time to another three children. On the other hand, she dabbled in the occult as an active Spiritualist and she was a member of a charitable order of women known as the Mission of the Flowers who dressed in robes and assumed the names of flowers. As a Lancaster, Maud was naturally known as Sister Red Rose.

Before Bill had become involved with airplanes, he and his brother had traveled out to Australia in 1914 as part of a Dominions Royal Commission that had subsequently been disbanded due to the war. He'd then worked as an electrician and a jackeroo on a sheep station before joining the Australian Light Horse cavalry, where he quickly became an expert horseman. However, when troops began to embark for overseas duty, he transferred to the engineers only to find trench warfare too claustrophobic and confining, and so in 1917 he managed a transfer to the Australian Flying Corps as a flight mechanic. But Bill was not there to stay on the ground, and by November he had trained as a fighter pilot and was commissioned as a second lieutenant. At this early point in his life, though, the sands in Bill's hourglass of luck apparently ran out, never to be entirely replenished. Caught in a snowstorm that reduced his visibility to a few feet, he was so badly injured when he flew into the ground that he was in a hospital for three months. Thereafter, Bill would be dogged by a series of accidents for the rest of his life, some of his making, some just fate, that would constantly turn potential triumphs into ashes in his mouth.

Although he flew occasional sorties, Bill's service in the Corps was terminated just prior to the Armistice in 1918 because of his medical unfitness. Somehow, he then managed to obtain a commission in the British Royal Air Force, probably because they were short of trained personnel, until he was demobilized with the honorary rank of captain in 1920. He promptly applied to rejoin the RAF, which he was able to do in April 1921. After five months, he was posted to India with 31 Squadron under the command of "Bert" Harris, later Sir Arthur "Bomber" Harris, Air Officer Commanding in Chief of the RAF during World War II, who would be responsible for the Dresden and Hamburg firestorm raids. Although he didn't tell Jessie this next part of his story until later, Bill had then married twenty-three-year-old war widow Annie Maud Mervyn-Columb, whom Bill and later Jessie always referred to as Kiki, just before he'd left for India. A beautiful, independent woman with a very determined streak, Kiki seems to have quickly become disillusioned with the brash and immature Bill. After little more than twelve months, he was posted back to England with Kiki and their infant daughter, possibly due to ill health or because he didn't fit in with the elite squadron or perhaps, as Jessie later suggested, because the scent of a scandal had clung to Kiki.

As his time in the air force drew to a close, Bill had gone out of his way to prove he was still a man. He volunteered for the new parachute jump training, was captain of the champion air force boxing team of 1925, and had proved he could still ride a bucking horse during the Tex Austin rodeo show at the newly opened Wembley Stadium in 1924, a feat that earned him the nickname of "Bronco" Lancaster. Since then, however, while Bill had been keeping up his flying hours, he was a civilian pilot without a steady job, tied to the ground by family life as other pilots flew into the air bound for adventure and unknown lands. Battling the tides of unpaid bills on a low income, reading in the newspapers of the financial rewards and fame of those who returned from afar in their planes, Bill had evidently dreamed of a flying adventure of his own to Australia, and now he'd found the listening ear of a pretty young woman who had her own dreams of escaping from the confines of her life.[22]

The more Bill talked about his dreams with Jessie, the more she warmed to him. However, she always refuted the myth about a love-at-first-sight meeting. "I liked him," she recalled, "I thought he was very nice but I wasn't in love with him or anything like that. We were just business partners."[23] Those earlier flights to Australia by Ross and Keith Smith, Parer and McIntosh, and Cobham had been made in large planes, some with crews, but nobody had yet succeeded flying that distance solo or with a co-pilot in a small light plane. At this point in time, though, Bill was flat broke and hadn't been able to persuade anyone, not even his wife or his father, to invest in an adventure that would cost serious money. Bill was a pilot, not a salesperson, but Jessie had earned her way to England by being a salesperson and she was good at it. She was the very person who could persuade potential investors to sponsor Bill's flight, for which there would be a lot of expenses, not just for fuel but also for landing fees, licenses, visas, servicing, spare parts and maps, and then there was the matter of finding an airplane. There would be some help available en route from airfields, RAF depots and bases across Europe and the Mediterranean to the Middle East and as far as India and Singapore, but after that they would be on their own over the remote islands of what was then the Dutch East Indies and then the Timor Sea until they reached Australia. Jessie could sense that Bill wasn't totally confident he could fly the entire distance solo and, in any case, he didn't seem capable of organizing the necessary funds. If it's left to Bill, she thought, this pioneering adventure will remain a dream until someone reaches Australia before him, and at the rate air records were being broken that was not going to be very long. Jessie liked a challenge, and so if this flight was ever going to happen there was nothing for it but to take matters into her own hands.[24]

For the next few days, Jessie gave the matter some thought, and by the time she and Bill met again she'd formed a plan of her own: she would help him organize the flight on the condition that she go with him. Bill, who hadn't seen this coming at all, was shocked. In his experience, an airplane cockpit was no place for a woman. Besides, Jessie had never been near an airplane, let alone piloted one. Maybe, Jessie retorted, but on the other hand, no woman had yet flown any equivalent distance. Think of the publicity, she insisted. They should take advantage of the current public fascination with pioneering women fliers, and having a woman as a flying partner would attract the kind of media attention money couldn't buy.[25] Once Jessie began to apply that carpet sweeper saleswoman technique, Bill was no match for her persuasiveness and, seeing he was about to lose this hand, he was forced to play his last card. He finally confessed that although he was currently living with his parents, he was actually a married man and a father, factors that seriously complicated their situation in those more morally conservative times.[26] It could prove to be very difficult seeking sponsorship from businesses and support from the conservative press as a man and a woman married to other partners who would be spending so much time together in such intimate circumstances. Jessie later claimed Bill attempted to clarify that he and Kiki weren't living together any longer with a story of a family scandal as a result of which he was being the better person by not seeking a divorce because of all the dirty laundry that would be displayed in court, leaving a trail of damage that Bill said he feared could ruin the children's lives.[27] While he did have a point about the terrible social stigma divorce carried at that time, Jessie could see that Bill was also using this as an excuse not to make a decision. She'd quickly realized that Bill typically preferred not to think things through too much; in fact, he'd often go to some lengths to avoid it, including bending the facts just enough to fit his story. He would later claim, for example, that he'd been given leave from the Royal Air Force for the flight to Australia, whereas by then he was no longer officially in it, and he would always refer to himself by his honorary title of Captain rather than use his actual service rank of Flight Lieutenant. As Jessie would discover in time, Lancaster was quite adept at telling "little white lies" that, while not complete fabrications, were what one might call Bill's version of the truth that usually put him in a better light. "I used to catch him out in little white lies," she recalled later, "although I don't know whether he was capable of any big deceit. I believed everything he told me about his wife."[28] For a while, this trait added to an aura of mystery around him, but eventually Jessie would grow tired of it. "I could never quite make out," Jessie said on another occasion, "which was absolutely the truth and what was said on the spur of the moment."[29] Years later, this habit of manipulating the truth would nearly cost him his life.

However, these conversations were taking place at the beginning of their friendship and so, having listened patiently to Bill's confession, Jessie thought the best way to deal with the elephant in the cockpit was for her to meet Kiki and ask, woman-to-woman, if she had any objections to Jessie flying to Australia with Bill. He, on the other hand, desperately tried to throw obstacles in her path, protesting that a meeting would be difficult to arrange, as Kiki wasn't living in London but was down south in Bournemouth, where she was the hostess of a private hotel, and that she'd be too busy and wouldn't be able to spare the time and, besides, he didn't have a car. Poor Bill had a lot to learn about Jessie, who was not easily diverted from a path to which she had put her mind. An opportunity for a flying adventure like this didn't fall into a woman's lap every day and Jessie, now determined to get into the air come what may, was not about to let Bill's inability to organize his domestic life come between her and her one big chance to take part in the adventure of a lifetime.

Jessie persuaded one of Bill's pilot friends to drive them down to Bournemouth in his open Morris Cowley. Of course, the two men sat up front under the canvas top while Jessie had to take the single "dickey" or "rumble" seat at the rear, despite the bitter cold and wind. Consequently, by the time they arrived at four in the morning an understandably furious Jessie had lost her good hat and looked a mess. Then, after only a few hours' sleep, she was woken by an immaculate and gracious Kiki, who pretended not to notice that Jessie was bleary-eyed and tousled. Jessie felt considerably disadvantaged and knew when she'd been outmaneuvered, but Kiki seemed friendly and soon the two women were getting along well. Later that morning, they all went out for a walk. When they stopped for a cigarette break, Jessie turned to Kiki and asked, "Now look, are you quite sure you don't mind if I go with Bill on this flight? He's agreed to let me go with him if I help him raise the money."

"My dear," Kiki merely replied, "I couldn't care less who he flies with or what he does, as long as he sends me some money."[30]

Having dealt with that problem, Jessie quickly proceeded to set the planning of their flight in motion. Making a list of Australian firms with offices in the city, she began to write letters and make calls inviting them to be involved in sponsorship of a historic airplane flight to Australia. Not only would this be the first and longest flight of its kind for a light plane, she would point out to people, but she would be both the first woman to make the trip by air from England to Australia and the first woman to fly that far. By August, she had gained the support of Sir Granville Ryrie, the Australian High Commissioner to England, who was only too happy to assure the public that Jessie and Bill could carry out their project successfully.[31] Wealthy Australian cattle baron Sir Sidney Kidman, who founded a company that is still the largest private landholder in the country, was also only too happy to endorse them, and he was followed by a number of other businessmen. Still, the immediate problem was that while they were certainly being promised money if they were successful, they had to put up a substantial sum just to get off the ground in the first place. "Millionaires who seem to find it difficult to spend their money wisely should remember airmen occasionally," Jessie observed later. "Flights simply eat up cash, and it is not only for the purchase of the machine that funds are required. There are thousands of things to be paid for. It's like a dressmaker's bill; you bargain on so much for the making of the frock, and then discover that extras have mounted up without your notice."[32] What they needed was more publicity and the *Daily Express* was quick to recognize the sales potential of a woman making such a flight. They bought the story on the spot with the condition that Jessie send them a cable every day from wherever they landed detailing their adventures. She and Bill already had plans for lecture tours after the flight was completed, and they were carrying at least one small movie camera to shoot film of their flight. Jessie's husband, Keith, was supportive and regularly sent her money, but when she wrote to her mother she promptly received a reply telegram along with the fare to return to Australia by ship immediately. Of course, Jessie just ignored it and put the money towards the expedition expenses.

Then Jessie and Bill paid a visit to Sir Sefton Brancker, Director of Civil Aviation since 1922 and an air pioneer in his own right, having been a pilot since 1913. Brancker had risen rapidly to senior air defense positions, culminating in his appointment to Air Vice-Marshall in 1919. In his current capacity as Director of Civil Aviation, he'd encouraged a number of English cities to build airports and airfields, and as chairman of the Royal Aero Club's Racing Committee he'd encouraged the formation of the Light Aero Club scheme in 1925, which had helped provide English clubs with new and improved light aircraft types such as the de Havilland Gipsy Moth and the Avro Avian. He was fascinated by their proposed flight and absolutely taken with Jessie. He suggested other contacts, listed the permits they would have to obtain in order to overfly various foreign countries and promised them his ongoing support.³³ Burbury supported them, too, with a 50 percent discount on their flying outfits. They would need maps, too, of course, and there was no better place for these in London than Stanford's in Whitehall. When Edward Stanford began his company in 1853, he was not only the sole map seller in London, but he was the only one who printed his own maps. In 1862 he published the first accurate map of the City, still on sale in the very same store they have occupied since 1900, the same store into which Jessie and Bill ventured for their aerial maps. There an RAF officer worked out their route with them, about 13,000 miles, and then that was converted by Stanford's into long strip maps divided into sections that folded up flat into a book.³⁴ Now, if one was setting out by car on a long trip by road, you just went into a gas station for a road map. However, when setting out on a flight around the world a long way off the beaten paths of established air transport, finding the right map with the right information on it takes time and you have to consult other travelers, for flight charts are essential, complex and valuable documents on which are drawn not only compass courses with their periodic changes, locations and distances to destinations but also features of local terrain, wind directions, local weather characteristics, critical altitudes and the ever-important emergency landing possibilities. Each section of the map was a leg of their flight, so rather than unfold a whole map they would only need to have one particular section open on their lap at a time. As Jessie and Bill finished each map book along their flight path, their plan was to post it to themselves in Australia to lighten the plane. Jessie paid the £40 for the maps out of her own savings, feeling that payment somehow sealed the deal between the two of them.³⁵

The maps were an ideal reason for Bill to introduce Jessie to his parents, with whom he was living and who were also looking after the two children, and so they went to visit Edward and Maud at their home on Crystal Palace Road, South London. Jessie found them understandably hostile to the idea that their son would be setting out into the unknown with a married woman as co-pilot, but she won them over with her openness and her willingness to share in the organizing and financing of the trip. In fact, they were soon so enthusiastic that they offered to contribute to the cost of purchasing the plane, but there were two conditions: Jessie and Bill would distribute pamphlets about the work of the Mission of Flowers wherever they went, and Jessie had to go with Maud to visit a spirit medium to confirm that she was the right person to fly with Bill. Despite the likely impact of a box of pamphlets on their tight weight restrictions, Jessie and Bill saw the advantages of having his parents as financial partners in the venture and willingly agreed to both conditions. Maud promptly whisked Jessie off to her favorite Spiritualist to ask his opinion. With some soft music playing in the background, he made them comfortable and then went off into a trance and invoked the spirits. They evidently gave him good news; when he awoke, he solemnly told Maud that the spirits had vouched for Jessie as Bill's flight partner. Then he caught Jessie's eye when Maud wasn't looking and gave her a broad wink.[36]

Now Jessie and Bill had to find a suitable airplane at a suitable price, which in their case was as cheap as possible. After discussions with various aircraft manufacturers, they managed to persuade A.V. Roe and Company to sell them a plane for about £300.[37] Although the public relations potential for Avro would be enormous if this record-breaking flight was successful, Jessie and Bill were still at this point unknown aviators who did not have enough celebrity status to be actually given a plane. "Avro" was an acronym for the name of the company founder, Alliott Vernon Roe, and they were the first company to be registered as an airplane manufacturer, in 1910. Later responsible for such classic aircraft of their time as the Anson, the Lancaster bomber and the Vulcan turbo-jet delta-wing bomber, the company still flies on as the oldest aircraft manufacturer in the world, now known as Avro International Aerospace, part of British Aerospace. Avro had quickly become famous for their Avro 504 fighter and training plane during World War I, and so by the time of Jessie and Bill's association with them they were already a well-known company. The Avian was still a relatively new airplane design, the prototype model having only been tested the previous year, but it was already proving popular in the civilian market as a light sports plane. Jessie and Bill's plane was the first Avro 594 Avian Mark III (R3/AV/125). Because of Jessie and Bill's specific needs, Avro had made a few modifications to this plane at their factory at Newton Heath, Manchester, including an extra fuel tank that provided a total 56-gallon capacity instead of the usual 35 gallons. The wings of the Avian folded back so that it was easier to handle on the ground by two people and could be fitted into small storage spaces. Despite reports that Avro was not sure the aircraft would be finished in time for the projected October departure date, it was completed and issued with Certificate of Airworthiness 1216 on 23 September 1927, having been registered with Certificate 1471 to A V Roe & Co. as G-EBTU.[38] To everyone's relief, Bill was able to personally take delivery of their airplane at the factory in early October and fly it to the Croydon airfield, from where they would depart on the flight to Australia. The plane was then registered in Bill's name with Certificate of Registration 1501 on 13 October.[39]

Bill had initially thought that the increasingly warmer and more humid weather into which they would be flying might affect the plane's wooden propeller, and so he'd fitted a spare inside the fuselage. Then the Fairey company made him an offer to use their new metal propeller, which would be much more durable. Bill accepted the offer, but when the new propeller was finally fitted and tested it wouldn't produce sufficient revolutions and had to be taken back to the factory to be recast. The wooden one was left with a friend in case they ever needed a spare at any time during their flight, and Jessie and Bill were left hoping they would be able to leave on Friday, 14 October. Meanwhile, they had also approached Sir Charles Wakefield and his firm of C.C. Wakefield & Co., inventors and producers of Castrol oil products, who were renowned for their sympathy to aviation ventures and who readily agreed to lend their name to the project and to not only supply oil but also lay it down in drums along the route.[40] British Petroleum and Shell also agreed to donate petrol.

Lancaster would pilot the plane from the rear cockpit, but it was fitted for dual control in anticipation that Jessie would co-pilot the plane from her position forward. There was just one small hitch: she hadn't flown yet. So, Bill arranged for his World War I pilot friend and long-distance record holder T. Neville Stack to give Jessie a lesson or two. Stack, who with fellow pilot Bernard Leete had flown two small De Havilland DH 60s from England to India in January, was only too happy to take Jessie up, and as soon as the weather was suitable they took off from Croydon in an open-cockpit three-seater. "I thought it was marvelous," Jessie remembered. "I simply adored it. I took to it absolutely. Never a qualm, never a moment of air-sickness, nothing. Bill always said I was a natural pilot."[41] However, she had yet to fly solo, and Jessie and Bill recounted their story of how that came about to Captain Jack Morris, head of the Morris Flying Service at Rodgers Airport in Philadelphia, and some friends in Jessie's suite at the Hotel Schenley in 1930 while they were waiting for the right weather for Jessie to take off on her record flight to Cuba. Pleased with her progress and the natural talent Jessie demonstrated at the controls of the *Red Rose* while he'd been instructing her, Bill had arranged to take her up at 8:30 one morning to give her some additional lessons in takeoff and landing. However, he overslept and arrived at the airfield two hours later only to find Jessie already in the air solo in the plane, soaring overhead in graceful circles. After a tense ten minutes for Bill while Jessie actually performed very well, she landed without any problems at all and then quickly took off again before he could climb in. Four or five landings and takeoffs later, she had proved her point that she was indeed a natural flier. When she finally taxied up to the hangar, Lancaster claimed, she told Bill that would teach him to keep his appointments.[42] Waiting at Croydon Airport on 11 October 1927, for the new metal Fairey propeller to be fitted to their rose-colored aircraft, Jessie must have looked as thrilled about the adventure ahead as she felt. Newspaper journalists observed she appeared "keen and determined. She laughs at the risks and possibilities of hardship and is impatient to start." Even Bill agreed that she was "cool, capable, and has a fine air sense."[43]

The following night, Sir Sefton Brancker gave a party for Jessie and Bill at Murray's London Club. During the evening, Brancker took Jessie to one side and asked her what she was taking by way of luggage. Jessie didn't know it, but this would be a crucial question because she would discover that her luggage, or to be more precise her lack of it, would be the subject of endless curiosity throughout her journey. Because the plane was small and their weight load vital to how much distance they could cover between refueling points, Jessie told Brancker, she only had room to take with her a small leather bag containing her toothbrush and a tube of toothpaste, a cake of soap, a comb and mirror, a small box of face powder, a clean pair of socks and a pair of silk stockings, one set of clean underwear, a clean shirt and a pair of shorts. In other words, all she could take with her on a journey from England to Australia that would last months was a bag weighing 7 pounds or 3.2 kilograms. Brancker nodded sagely and then advised her to take along two other items: an evening frock of light chiffon and a pair of lightweight evening shoes. Jessie looked politely incredulous. Believe it or not, Sir Sefton explained, she and Bill would be invited to all kinds of celebratory functions along the way and she would feel embarrassed if she had to attend them in the clothes in which she'd been flying. Somewhat skeptical, Jessie nevertheless thought she'd better take the advice and rushed into London the next morning to buy the evening frock and a pair of black satin slippers. Sir Sefton was right; as it turned out, Jessie needed that dress and shoes nearly every night and they lasted the entire trip.[44] Many years later, Elsa Schiaparelli would make an entire trousseau for her "Constellation" collection of 1947, including coat, six dresses and three hats, that fitted into a specially designed Constellation bag all of which weighed less than 10 pounds, but early women fliers such as Jessie with their minimalist luggage were way ahead of her in principle, if not in looks. Amelia Earhart would often take little more than toothbrush, toothpaste and a comb and even in 1934 Freda Thompson, the first Australian woman to fly solo from England to Australia, only took a day dress, a night dress, a change of underwear and another pair of shoes in a tiny case crammed in alongside the auxiliary fuel tank.[45]

Jessie and Bill's final hurdle proved to be more paperwork. Because of the high risk involved, they couldn't find anyone who would insure the plane for the trip. They certainly didn't feel like taking off uninsured on the thirteenth, and so they postponed their departure until the following day. Perhaps once they'd actually left Croydon, an insurance company might see that they really intended to make the trip. Bill's father agreed to keep trying to organize it while Bill and Jessie concentrated on getting their plane ready to fly to the other side of the world. All they needed now was fine weather for the departure day, but when a crowd of people gathered to say farewell to Jessie and Bill on the chilly morning of Friday, 14 October 1927, they had to find each other inside a fog bank.

Two

Into the Air

At least it was a large crowd. After all, their flight plans had been the subject of newspaper reports and speculation for the last six weeks. Expert opinion seemed equally divided on whether they'd make it to Australia or not; while a few had succeeded in flying the distance, many had failed. Of those six planes that had attempted to reach Australia from England during 1919-20 in response to an Australian government prize money offer, only two had reached Australia intact. The remaining four all crashed, in two cases killing both pilots and co-pilots. Those six planes were all large aircraft for their day. No solo aviator had yet completed the entire England to Australia route in any type of aircraft, let alone a small, light one. However, Capt. Dennis Rooke had given it a good try earlier that year. Leaving the Croydon airfield on 24 May in a DH60 Gipsy Moth biplane, he'd made good time until while landing at dusk at Karachi he'd mistaken a short cricket field for the longer airfield and damaged his plane. By the time repairs could have been carried out, the monsoon season would have arrived and flying a light, open aircraft in those conditions was too dangerous to even consider. Rooke had to concede defeat and that left the door open for Jessie and Bill. Even today, in an era of satellite-assisted navigation and communication, you would give serious thought to flying an aircraft the size of an Avro Avian over the distance that Jessie and Bill were considering, yet in 1927 they contemplated it knowing they would be landing in places where there were no airfields and where the local people had seen few Europeans, if any, and had probably never seen a plane. People could die out there, as evidenced by Elliot's death during Cobham's flight the previous year.

More important for Jessie and Bill's publicity, however, was that no woman had yet traveled any equivalent distance in the air, certainly not to Australia, although they were certainly attempting it. Austrian actress Lilli Dillenz, for example, had left Lisbon the same day that Jessie and Bill left England, flying as a passenger with two German pilots and two crew in a modified trimotor Junkers headed for the Azores in an attempt to be the first woman across the Atlantic in the air. Once there, however, they were unable to persuade their plane to take to the air because of the weight of the necessary fuel load. The air could also be just as dangerous an environment for women as for men, for that matter. During the preceding week to Jessie and Bill's departure, Ruth Elder had been attempting to be the first woman across the Atlantic with her flight instructor Capt. George Haldeman in their Stinson Detroiter *American Girl*, but just the day before they'd had to be rescued by the steamer *Barendrecht* after their plane was forced down at sea, only 300 miles from their goal.[1] Elder was lucky: two other well-known women pilots perished at around this same time during long-distance record attempts. Student pilot Mildred Doran, aged only twenty-two, was a crew member aboard the *Miss Doran* when that plane vanished at sea during the deadly Dole Air Race to Hawaii in August 1927, during which ten aviators lost their lives. Frances Grayson would be lost aboard *The Dawn* when that airplane and crew disappeared during their transatlantic attempt in December 1927.[2] The following year Elsie Mackay, also known as actress Poppy Wyndham and one of the first women in England to gain a pilot's license, along with her pilot, Captain Walter Hinchliffe, would disappear off the coast of Ireland in her Stinson Detroiter *Endeavour* while attempting the same journey from England. From a global perspective, Jessie and Bill's flight was just one of a number of attempts on distance records taking place at about this time to various destinations, keeping the public avidly reading their newspapers. Charles Kingsford Smith was meticulously planning his flight from America to Australia, while Captain Frederick Giles had just taken off from San Francisco for New Zealand on his transpacific attempt and the two French aviators Dieudonne Costes and Joseph Le Brix had just left Paris on their 35,672-mile flight around the world.[3] But it was the presence of Jessie that made the flight of the *Red Rose* stand out from the contemporary crowd of eager aviation record breakers: this would be the longest flight by any woman, passenger or pilot. This was Jessie and Bill's major marketing ploy, and to a large extent it worked. Without Jessie, Lancaster would have just been another one of any number of brave or brash pioneer aviators taking off into some far blue yonder that year. With Jessie, this flight would be the first of its kind and she would become the first woman to fly into another hemisphere on the other side of the world.

All morning, the crowd at Croydon waited patiently for the weather to lift. Jessie posed with Bill by their plane for photographs, in between making sure that everything was stowed on board as compactly as possible. Their Avro Avian open-cockpit biplane was only a small aircraft with very limited storage capacity for fuel and luggage. With two people on board such a light aircraft, weight would be a critical factor in just how far that fuel would take them, and so everything that could be squeezed into the small luggage compartment behind Bill's rear cockpit had to be considered in terms of necessity: just some oil, a few essential spare parts and tools, emergency rations, maps, a firearm and ammunition, and their personal items. Jessie would constantly find herself an object of wonder to the media and to other women during the trip because her luggage for a journey to the other side of the world consisted of just one small leather bag. When she left England, she was wearing brogues and jodhpurs, but once into the tropics she exchanged these for a white shirt and the shorts for which she would become famous. She didn't use makeup and kept her hair cut in a short bob parted in the middle.

Finally, in the early afternoon the weather started to clear and there was a bustle of last-minute preparations and good-byes. The Australian High Commissioner, Sir Granville Ryrie, along with Lady Ryrie and their daughter, were there to wish them well. Bill and Jessie officially accepted a letter from Sir Granville to the Australian Prime Minister, Stanley Bruce, and another one from Sir Sefton Brancker. Colonel Ivo Andrews from the Air Ministry was there to offer an official farewell, and standing back in the crowd was a young test pilot for Avro by the name of Bert Hinkler who had already made a record flight in his earlier-model Avian from this airport to Riga in ten hours. Lady Ryrie christened the tiny biplane *Red Rose* by scattering a bowl of red rose petals over it, officially naming it both after Bill's mother and in memory of his Lancastrian heritage. Jessie was kissed good-bye by Mr. and Mrs. Lancaster and Kiki and handed a bunch of white heather for luck. Newspaper journalists were already taking note of Jessie's choice of clothes, a rather patronizing but fascinating media obsession that persisted throughout the duration of the flight and into Australia. Today, they observed, she was wearing knickerbockers, a striped sweater, thick stockings and brogue shoes. However, to give the Sydney *Sunday Times* journalist credit, they also called her a "heroic woman" who was "credited with being the driving force of the enterprise and [who] has handled all the business details." Jessie simply declared she'd always wanted to be the first woman to fly to Australia and that she hoped to realize her life's ambition.[4] Then, just before two-thirty, she and Bill put on their helmets and flight jackets, over which Bill shrugged on a full-length heavy overcoat and Jessie donned a life jacket, and they climbed into their respective cockpits that they would occupy for many weeks. Bill's wife, Kiki, who hadn't been living with him in all the time Jessie had known him, had decided to travel from Bournemouth to take part in the farewell for the sake of good publicity, and so she was there to lift up one of their daughters for a last kiss good-bye from her father. Then the prop was spun, the engine started, the chocks whipped away from the wheels and they taxied into place on the runway. To the cheers and calls of the huge crowd, they raced down the tarmac and lifted off into a wide, blue sky.

A short time later, they touched down briefly at Lympne Aerodrome in

Kent, home for the last four years to light aircraft trials and races and the starting point for a number of long-distance record flights. While the plane was being inspected by Customs, Bill telephoned his father about insurance, but it had still not been forthcoming and so, despite his parents' protests, they left the shores of England uninsured. They flew over Folkestone and out over the Channel, only to encounter more fog. Conditions became so bad as they struggled towards the coast of France that they lost sight of the water and had no idea they'd crossed land until darkening skies forced them to look for a landing place near Abbeville, not far from the mouth of the Somme River northwest of Paris. While Bill studied his charts, Jessie took the controls and became more confident handling the plane until Bill took over to land it. They came down in a field a short distance from the Abbeville airfield, and after walking over there with their bags they eventually found a place to stay. As Jessie laid her head back on the pillow that first night, her mind was a confusion of hopes, desires and ambition to pilot her own plane one day.[5]

The next morning they flew on to Paris, passing over the great unfinished Gothic cathedral of Beauvais and a French airliner that had likewise been forced down by the weather. Touching down at Le Bourget in Paris, they were met by pilots of Imperial Airways whose route they would follow fairly closely all the way to India. After they spent some time checking the plane, filling the petrol tanks and making sure all was in order, it was time for a night out in Paris with the pilots. After some excellent food and wine, they all took their shoes off and paddled in the fountains outside the Folies Bergères. That night provided some wonderfully warm and joyous memories for Jessie to draw on as comfort during the rougher times ahead.

They received news the following day that the British Aviation Insurance Company had finally agreed to insure them. Bill's plan had worked; their departure had provided them with leverage. From Paris they flew south, supposedly into warmer temperatures, yet winter appeared to be pulled along behind their plane. To Jessie it seemed as if "all the fogs in the world were dogging our propeller."[6] In appalling conditions, they were forced to fly by their compass through dense cloud and driving rain, making barely 60 miles an hour and sometimes dropping as low as 100 feet, until the cloud cover parted enough for them to slide through and below it to land at Lyons. In those days, long before the invention of radar, airplanes didn't carry wireless equipment or night-flying instruments. They weren't referred to as "ships" for no reason; they were literally ships of the air, navigated and steered in similar fashion to ships at sea. The pilot plotted their location using a map, a compass and often a sextant, but these were fine weather instruments when used in open cockpits. If you were flying solo, you could hold the plane steady while using navigation tools by pinning the control stick between your knees, thus providing one rather tongue-in-cheek explanation of how it came to be known as a "joy" stick. More often, you navigated by simply lifting yourself slightly in the cockpit, or dipping the wing of the plane, and looking over the side to read the local geography. You followed rivers, trails, roads and, in particular, railroad tracks, which usually traveled in the straightest line between places and were thus often referred to as the "iron compass." If the cloud was too thick to see the ground, then it was back to your onboard compass and an excellent sense of balance that would keep the wings level and the nose up when you couldn't see the horizon. In these circumstances, one literally flew "by the seat of your pants."

Fortunately, by the next day the weather had cleared and Bill and Jessie flew into brilliant sunlight for the first time since they had left England on their way to Marseilles. Even so, as they flew down the Rhone Valley, passing hills crowned with castles and churches, they had to cope with a lot of thermals. When the *Red Rose* suddenly hit an air pocket and dropped 300 feet without warning, Jessie shot straight up from her seat and hit her head on the center wing section before falling back into the cockpit again, slightly stunned. It was an abrupt reminder of the value of her seat belt. More than one unsecured aviator had fallen to their deaths when violent weather had pitched their plane on its nose or flipped it over. Arriving at Marseilles, they were welcomed by another group of aviators, who were only too happy to help them with their aircraft maintenance. That turbulence had loosened some strut bolts and the flying wires between top and bottom planes had become slack. The airfield was situated a long way from town, and so when they were finished for the night the crowd of hungry aviators had to travel some distance by train and bus before finding a place where they could eat sardines and onions and hunks of bread and cheese while drinking local wine. Time and time again throughout their journey, Jessie and Bill would experience the close bonding, mutual respect and companionship of a fellowship of the air that had no national or cultural boundaries. Although one of the few women then aloft, Jessie was always treated equally by her peers as another flier; it was civilians and the media who typically saw her as just "a little woman."

On 18 October, they left Marseilles for Marignano and Pisa. They flew over Cannes and Nice, following the coastline and enjoying the blue of the Mediterranean, although keeping a watchful eye on the hills that came right down to the water's edge as they flew on into Italy. They happily greeted Pisa's Leaning Tower by circling it twice, landing at the airfield to be officially greeted by the Italian Air Force. But by late afternoon it was raining, and it was still wet and gloomy the following morning. Bill watched the weather carefully, and as soon as there was a break in the cloud cover, about noon, they took off for Rome. It proved to be only a brief respite, though, and low clouds and freezing rain quickly forced them to land at Vaccina, where air force officers rushed them to the mess to warm up.

However, Jessie and Bill were anxious to reach Rome before the light was gone, and so they braved the weather and took off again, reaching their goal right at sunset. Flying along the Tiber as the sun glinted from the surface, Jessie thought the river's many bridges looked like clips dividing a silver ribbon into sections. The Caesars, she marveled, would never have dreamed that someday people would approach their fabled city on its seven hills from the air like this. Jessie and Bill's tight schedule pushed them back into the air early the next morning, though, with no time for sightseeing until they landed in Naples. Here they were finally able to take a break, accepting an invitation to a tour of the city with the president of the local Aero Club. They returned the favor by taking him in turn on a tour of his city from the air. When they took off on the way to Catania, Jessie and Bill circled Mount Vesuvius, from which, she mused, smoke curled "as from an evil pot."[7]

At Catania they were ceremoniously presented by Signor Balboa, Minister for Air, with an official message of congratulations and best wishes for a successful trip from Mussolini. Carried away with enthusiasm, the local air force personnel wrote "Viva Mussolini" all over the plane, but then as Jessie was sitting in the rear cockpit pumping up petrol with a hand pump she crushed her knuckles on the side of the plane and bled profusely everywhere, all over the writing. The Italians rushed to the rescue with iodine just as Signor Balboa came over to bid farewell. He shook hands with Bill and then, before a horrified Jessie could stop him, he gallantly bent down and kissed her hand. To her horror, when he lifted his face it was covered in blood and iodine! Handkerchiefs were hurriedly produced and the Minister's face was quickly scrubbed. Everyone held their breath until he burst out laughing, cheerfully seeing the funny side of it all, and they parted good friends as the tiny plane lifted off and headed out over the sea towards Malta.[8]

Although the Royal Air Force based there had promised an escort into Malta, they were cautious about their weather and so didn't fly out. Consequently, Jessie and Bill had to find their way to the local airfield by themselves. They took another break on the island and ended up indulging in the local life so much that, by the time the weekend was over, Jessie figured she would never look at a St. John Ambulance nurse again without thinking of Malta.[9] When they finally took off, the Royal Air Force made up for their previous lack of attention by laying on three seaplanes to provide an escort for about a hundred miles out across the Mediterranean towards Tripoli, while the Royal Navy made sure one of their cruisers was also on station at about the halfway mark. For three and a half hours, Jessie and Bill flew over water with no land in sight at all; their only distraction was buzzing the decks of a large passenger ship and getting a laugh out of some startled faces. As they finally flew over the North African coast, remarkably within two miles of their intended landfall point, Jessie Miller became the first woman to have crossed the Mediterranean Sea from Italy to North Africa in the air.

Jessie was presented with a huge bouquet of flowers at Tripoli by the Italian officers there, and they were honored at a tea dance that afternoon and at a British Consulate dinner that night. They had planned to fly along the coast to Benghazi, but the Italians warned them the Arabs in that area were a little hostile because a road was being built between Tripoli and Benghazi and the local tribes were unhappy about it going through their land. They might assume the plane was Italian and take a few shots at it, so the Italians advised Jessie and Bill to fly a Union Jack to clarify their nationality. The next morning, a silk automobile flag was duly found and hung from a wing strut and then, with the Italian officers and British Consul looking on, Jessie had to squeeze herself and the huge bouquet into her tiny cockpit, politely waiting until they were out to sea before ditching it over the side so that she had some room to breathe.

Stopping off to inspect some Roman ruins along the way at Homs, they continued to fly along the coast towards Sirte while heeding the warnings and remaining at about 4,000 feet. It may have been just as well they did. Looking over the side of her cockpit, Jessie could see clusters of tents and people far below and then she saw four puffs of smoke, as if someone could have been firing at them. She was still mulling over what she had seen when she was startled by a shout from Bill. Turning quickly to look ahead, Jessie couldn't believe her eyes. There in front of her, where moments before there had been clear sky, was a towering wall of flying sand! She just had time to push her goggles firmly down over her eyes when they were enveloped in a blinding storm carried on a 45 mile-per-hour wind. In an instant, sand was gritting their teeth, clogging their ears and sifting down the backs of their necks. Conditions became worse by the minute: the floor of Jessie's cockpit quickly became a beach and then the carburetor began to choke. By now they were down to 150 feet and were virtually being driven backward by the storm. It was either find a landing spot now or risk being blown out of the sky and into the ground. Thinking they had flown over Sirte in the storm, they turned back to descend to the first town they saw, only to discover on landing that Sirte was still an hour ahead of them. Now there was nothing they could do except shelter behind the plane as best they could from the stinging sand while they hung desperately on to the wings for what seemed hours to prevent the wind flipping it over. Although it seemed it would go on forever, the storm eventually blew over them and they were able to shake off the sand and take off again, finally landing at Sirte right at dusk.

At this time, Sirte was a major Italian Air Force base.[10] Jesse and Bill had brought with them newspapers from Rome that carried their story across the front pages in large letters and these served as a great introduction for two people who spoke no Italian. As far as anyone could remember, Jesse was the first and only Western woman to set foot in the town, and the bachelor base commander was a little puzzled at first as to how to accommodate her. In the end, while Bill of course shared the commandant's quarters, Jessie was given a stone-floored room in the soldiers' barracks furnished with an iron bedstead and a small card table, in which a number of cockroach families disputed occupancy rights. Still, Jessie graciously communicated her thanks in sign language to the Commanding Officer when she and Bill were treated to a wonderful dinner that night, and having decided he quite liked Jessie, the CO presented her in return with an aerial photograph of the air base. After dinner, the aviators and the officers sat around on the veranda drinking and smoking. Even at night, it was incredibly hot and the sand-filled wind was still blowing so hard that Jessie had difficulty walking into it on her way back to bed. Not having a book to read to help her sleep under her mosquito net while she endured the heat and the cockroaches, Jessie studied the photo the CO had given her. To her amazement, she realized the photo revealed a series of underground bunkers around the perimeter of the airfield large enough to conceal planes and oil and petrol dumps and that, consequently, she should show it to the British Consul in Benghazi the moment they landed there. But when she emerged from her room at first light, she was met by three officers who quietly begged her to return the photo before anyone found out what the CO had done and had him shot. Jessie really didn't want anything that drastic to happen to him on her account and so she gave it back to them, having already memorized the layout anyway. Then, after some strong, sweet black coffee, Jessie and Bill farewelled the Italian Air Force and pointed the nose of the *Red Rose* towards Benghazi.[11]

They flew east across the azure waters of the Gulf of Sirte, before descending to Benghazi late in the afternoon, where they were met by the British Consul, to whom Jessie duly passed on her information. The weather was calm that night, but by the time they reached Sallum the next morning, just over the border into Egypt, the wind blowing off the land had become so strong that it became increasingly difficult to make an approach to this town sited on top of sheer cliffs. To make matters worse, the airfield there wasn't marked and they couldn't find it. Even while they circled repeatedly, trying to find a hidden landing area while being thrown around the sky by violent updrafts, Jessie couldn't help but be struck by the beauty of the seascape beneath her where a vivid blue sea pounded against the cliffs. It was all lost on Bill, however, who was becoming desperate to find a landing spot as he watched his petrol gauge quickly sinking towards empty. In the end, he came up with the dubious idea of using some smoke bombs given to them by the Italians to mark a landing zone and indicate the wind direction, and he promptly proceeded to drop a series of them over the side one after the other. They did the job far too well. When the smoke cleared a little as all the plumes blew in the same direction, Jessie and Bill could see troops and civilians on the ground running in all directions like a disturbed anthill because, as the aviators found out later, they thought the local fort was being bombed. Then, as the *Red Rose* came in to land, the tail skid struck some hidden rocks and suddenly snapped, throwing the plane sideways and immobilizing it. When the dust settled, they were surrounded by pointed rifles and a very annoyed commanding officer demanding in excellent English to know just who they were and what they thought they were doing. Bill immediately became the gallant English airman, dismissing all concerns with a wave of his hand, introducing both of them and asking if by any chance there was any food to spare on the base.[12]

What is really remarkable in this day and age about Jessie and Bill's great aerial journey into the unknown is that they, much like British explorers of the earlier nineteenth century before them, believed and expected they would be extended hospitality and help wherever they went because they were white and British. It's the equivalent of an ice skater having absolute faith the ice will always support them or of a mountain climber believing their rope will always hold should they fall. Jessie might have been Australian, but at this time there was really little cultural difference between an Australian and an Englishman. Australians still largely thought of themselves as being British, tried to speak and act as if they were British, studied British history in school to the detriment of their own, sang the British National Anthem, used the British flag and waxed childishly nostalgic about the Mother Country that they continued to visit every chance they had. So, secure in the faith of the Empire, Jessie and Bill set off across half a world about which they knew virtually nothing, knowing that the people of the Empire would support them. They carried no foreign currency (in fact, they carried little money at all) and knew no language apart from English and virtually nothing about the cultures they would meet, even if they knew who they were. Remarkably, none of this appears to have bothered them in the slightest. They were totally confident that because of their white skin and British passports they would be protected and looked after wherever they went as part of the Empire cultural family that held them firmly within its embrace.

So, they were not surprised at all when the ranks of encircling Italians at Benghazi suddenly parted and, like Moses through the Red Sea, up strode Pasha Major-General Sir Charlton Spinks, who just three years previously had achieved the unique rank for an Englishman of Sirdar, or Commander in Chief, of the Egyptian Army. As it happened, he was just passing through Benghazi on his annual inspection of troops, along with Major Bailey, Governor of the Western Desert, and their respective staff, and there was much shaking of hands and exclamations of "jolly good show" and, before they knew it, our intrepid aeronauts were being invited to the regimental banquet being held that night under a large marquee for some sixty officers in honor of the Pasha's visit, hosted by Lieutenant-Colonel Ekainaham Hussein Kamel. After the band had played "God Save the King" and they'd been escorted in, Jessie found herself seated between Spinks and Bailey for a seventeen-course marathon of food. Seven meat courses alone were wheeled in during the evening, including an entire sheep from which diners hacked off their own portions. "I lost my hunger long before it was finished," Jessie wrote. "The very sight of a whole sheep is apt to take the keenest edge off one's appetite." After it was all over, Jessie was only too glad to sink into Spinks' personal bed that he'd kindly vacated for her and sleep it all off.[13]

They landed at Aboukir (now Abu Qir) on the outskirts of Alexandria on 28 October, followed onto the airfield by another remarkable air expedition. Eleven days previously, four twin-engine Royal Air Force Supermarine Southampton II flying boats, known as the Far East Flight, had left Plymouth in England on a 28,000-mile aerial voyage that would take them to India, Singapore, Australia, Hong Kong, and then back to Singapore in December 1928. Led by Group Captain Henry M. Cave-Brown-Cave, the expedition was designed to test the long-range capacity and durability of the self-contained, duralumin-skinned flying boats on which each crew of four could live for several days at a time. This unique flight of aircraft was a spectacular sight as they came in to land at Aboukir, and once down on the airfield Cave-Brown-Cave came over to greet Jessie and Bill and to discuss with them each other's aircraft and routes. As they were all heading in the same direction, they made plans to eventually meet up again at Karachi.

From Aboukir, Jessie and Bill flew on to Heliopolis on what was then the outskirts of Cairo. This new city, not the ancient one long buried under northern Cairo, had been established in 1907 as a city of luxury and leisure in the desert outside of Cairo by the European entrepreneur Baron Empain. By the time Jessie and Bill arrived, there was a Royal Air Force base on the outskirts of the city. When Squadron Leader Victor Erskine Lindop, the Commanding Officer at Heliopolis, learned that they were about to fly out into the desert in a plane too small to carry any extra water or emergency rations, he made them wait until they could be escorted by an RAF Vickers Victoria that would leave on Monday for Baghdad. Given an unexpected weekend stopover in Cairo, Jessie at last had the chance to be an excited tourist and to spend two days sightseeing, including exploring and climbing the pyramids. She and Bill socialized as well, of course. Bill had a chance to meet some of his old pilot friends from when they had served overseas together, and there were dinners to attend at night. By now, it had become something of a tradition on this flight for people to sign their names on the lower wings of the *Red Rose*, and more were added before Jessie and Bill's departure.[14] So, at seven o'clock on Monday morning, they took off in company with the twin-engine Vickers Victoria, a freight and troop carrier that flew a regular supply run for the RAF bases between Cairo and Baghdad. As the two planes turned northeast, they flew over the Nile. "The sails of the strange craft on the river looked like little bits of paper blown by the breeze," Jessie wrote, "and spread beneath us was a living picture of the maps we carried in the cockpit."[15]

The Mediterranean coastline disappeared into the distance as they flew out across the desert towards Ziza (now Al Jiza), a few miles south of Amman in Jordan. As they droned on into the morning Jessie observed that, instead of glittering in the sun as she expected, the vast expanse of sand beneath her reflected the blue of the sky as if measuring the space around her in the air with its own emptiness.[16] They flew for hours over a universe of sand, seeing no oasis, no camels, no tribes dressed in flowing robes, nothing but sand. The sun grew hotter and the slight breeze stirred by their passage seemed to gust from the mouth of some blast furnace. When the two planes landed at Ziza the group discovered, after much searching of both planes, that in the hurry to leave no one had actually packed anything to eat, so two of the RAF officers flew across to the nearby base at Amman, eventually returning with bottles of beer and with tins of bully beef that they heated up on a handy saucepan lid before dishing it out onto newspaper. Then it was into the air again, northeast towards Persia (now Iraq) and the fabled minarets of Baghdad, crossing territory that is remote even today. Jessie watched the landscape beneath her gradually change from sand to flat, barren, basalt country folded with hills and valleys, crowded with great boulders that she thought looked black and sinister. In 1921, members of the Royal Air Force had actually dug a furrow all the way across this desert from Palestine to Baghdad to direct its planes there; six years later, Imperial Airways basically used the same line when they opened the Cairo–Baghdad–Basra section of its England to India route, and this line was what the two planes were now following.

By now, Jessie and Bill had worked out a system of communication between cockpits. Sitting out in the open, fully exposed to wind and engine noise with no such things as microphones or headsets enabling clear voice communication, you either shrieked monosyllables very loudly, used sign language, or wrote notes and passed them over while hoping they were not torn out of your hands by the wind and lost. About thirty minutes outside of the Rutba Wells, their halfway point, Bill handed across a note to Jessie asking, "What are you doing with your feet? Are you thumping them on the floor?" She wrote back indignantly she was doing no such thing. "Well, something is happening!" Bill wrote back. Right about then, she began to smell burning, not something you want to smell a few thousand feet in the air inside a very inflammable aircraft made of wood and fabric and filled with fuel. They promptly descended towards Rutba, breathing a sigh of relief as they landed without having burst into flame. Leaping out to examine the engine, they discovered the driving sprocket on their impulse starter magneto had sheared and was smoldering energetically in the heat.[17]

Fortunately, Jessie and Bill had both company and help for this emergency. The Rutba Wells were the only water for a long distance in any direction and consequently served as a meeting place for nomad tribes as well as a stopover point for Imperial Airways. An impressive gray fortification overlooked the site, occupied at that particular time by a British armored car patrol and an Imperial Airways representative. Before starting in on the engine repairs, Jessie and Bill shared some more bully beef, biscuits and tea with the Vickers crew who had landed with them before waving them goodbye. The faster RAF plane had been forced to travel much slower to escort the Avro Avian and now had to resume its usual schedule. That night, Jessie and Bill were the guests of the armored patrol personnel for dinner under the stars. Seated on upturned kerosene tins around a roaring fire, they had stewed gazelle as a main course, served on a limited number of saucepan lids that had to be passed on to the next person when one diner had finished, as were the only two mugs for tea. Way out in that wilderness beyond the fire's glow, wild animals howled around them in a night as black as pitch.

They had little rest that night. In the early hours of the morning, a sandstorm blew up that threatened to overturn the plane where it sat outside the walls, so Jessie ventured out into the blasting sand and howling wind to check on its welfare. As she approached the plane, she heard a voice shouting in Arabic but of course had no idea what was being said until suddenly the muzzle of a rifle came up against her nose. After she had urgently identified herself to the native guard in the short, flickering flame of a match, she made a note to herself that it might be a good idea to learn some basic Arabic as soon as possible. Between them, Jessie and the guard only just managed to make the plane secure before they were suddenly struggling in heavy rain instead of sand. Folding the wings, they wheeled the *Red Rose* through the gate inside the walls of the fort with only a couple of inches to spare on either side. No one had ever managed to fit an airplane inside the Rutba fort, and so in the light of the day there was a considerable amount of headshaking, inspection and photographing while Jessie and Bill worked away busily.

It was well into the morning before the sand-clogged engine could be persuaded to turn over and stay running long enough to take off. Always the man of action before thought, Bill insisted on taking the *Red Rose* into the air before Jessie considered the engine could be trusted, and her female intuition was spot on: the engine started cutting out after only a few minutes. Four times it died and four times Bill managed to restart it before he ran out of air, but the fifth time it stubbornly refused to fire until they were almost into the ground. By then, Bill was taking more heat from Jessie than the desert sun, and he grudgingly condescended to land briefly on a desert track near Ramadi to inspect the engine, but, still insisting they could make it to Baghdad on schedule, he quickly took off again.

Flying over the Tigris in the middle of the afternoon of 3 November, Jessie thought the river looked like an angry snake squirming its way across the countryside, which was pretty much how she felt. By then it was raining hard and she was cold, wet and tired.[18] When they attempted to land at the Royal Air Force airfield at Hinaidi, just outside Baghdad, the runway was saturated and dangerously muddy, but after some careful maneuvering they landed safely and pushed the plane into a hangar out of the weather. By the following day, however, the airfield was completely underwater and any attempt to use it was out of the question. They had little choice but to sit around and wait until the rain stopped and the ground surface dried out enough to take the plane's weight, light though it was. Jessie had been looking forward to the Baghdad of Aladdin and *A Thousand and One Nights*, but this was the gloomy, poverty-stricken Baghdad of 1927, a "city of dead romance" that was now like "a faded tapestry with only here and there some of the original color remaining," she wrote. The haphazard narrow streets weaving their way among acres of flat roofs along the banks of the muddy, sluggish Tigris

The *Red Rose* at the Hinaidi Royal Air Force Base, outside of Baghdad, early November 1927, on the way to Australia (courtesy John Pollock).

seemed to her to be "trying to live up to a past reputation and failing dismally in the attempt." Having thought of herself as arriving in a modern version of Sinbad's flying carpet, she concluded that it was like having your faith in a fairy tale destroyed.[19] Nevertheless, despite her evident disappointment in the city, Jessie would always treasure a gold elephant-hair ring presented to her by an RAF officer there to wish her luck. While on the one hand Jessie would always swear she was not superstitious, although born on Friday the thirteenth, she never went anywhere again without that ring, nor, incidentally, would she ever light three cigarettes with one match.[20]

Five days dragged by until the airfield was dry enough to support their weight and they could put air under their wings, turning south now in the direction of Basrah at the head of the Persian Gulf. Before them, a great expanse of marsh and swampland spread out like a scroll being unrolled from the rim of the world by an unseen hand, flooded after a week of rain. The two great rivers, Tigris and Euphrates, seemed to ooze rather than flow through the landscape with their load of silt. Evocative names to read about, Jessie wrote, but "their sullenness is heavy with evil omen," and so it proved to be.[21] She and Bill hadn't been in the air long before it became obvious that, slowed down by the weather, they were never going to make Sheibah, the Basrah airfield, by sundown. They detoured to Wasiriya, but as they flew low over it Jessie could see that it, too, was nothing but a sea of mud. If they tried to land in that, the wheels would sink and they'd nose straight into the ground. Circling back, they found the railroad tracks and followed them down to Ur Junction, 100 miles from Basrah and the only good landing ground anywhere near. It was just as well; the engine was coughing badly and it would be much too risky to fly any farther. They hadn't been on the ground very long, though, before a pair of RAF aircraft on patrol joined them and the night became a lot less lonely for them all.

In the light of the next morning it was clear the magnetos needed repairing, and so the patrol took them on to Basrah and then returned to drop a message. As they ran to pick up the canister, in which was a note that read the Base CO was about to arrive with their parts and would then escort them personally into Basrah, they left the plane briefly unguarded. By the time they returned, it was surrounded by local people and the aviators' precious strip map book was gone. Now in 1927 this was a potential disaster, the modern equivalent of which would be having the GPS stolen. With no knowledge of the area, no language skills and no radio, the aviators literally did not know where they were without their maps. Away out here in the desert, they couldn't stop at the nearest gas station and buy some more. So, Bill had to go in search of the nearest telegraph office from which to send a wire off to the Hinaidi RAF base, who then telegraphed ahead to Karachi to have maps waiting for them there. In the meantime, still surrounded by some dubious-looking robed figures sitting in the sand, Jessie guarded the plane with a pistol in her hand. None too soon, the CO arrived with the much-needed magnetos, but, despite the new parts, the engine still refused to run smoothly. Still, it was at least marginally cooler in the air than in the hellish heat on the ground and so they staggered on to Basrah escorted by the RAF.

There they heard that almost the entire village at Ur, including women and children, had been imprisoned for the theft of the maps, but the thief was never identified. This kind of treatment of local inhabitants was not unusual in those colonial times. Although Iraq had been a British mandate since the Turks had been defeated there during World War I, there had been a major revolt a few years previously in 1920 and for four months rebels had occupied towns and killed British troops in an effort to end the occupation. The British considered this was rather ungrateful, considering the efforts they had made to introduce civilization, and so Winston Churchill's civilized reply had been to bomb villages, arguably at times using poison gas, an action that he argued had an excellent moral effect on the inhabitants. As if that weren't enough, hundreds of people were imprisoned under a harsh punishment regime. Some 10,000 Iraqis probably lost their lives, so one can understand that both sides were still somewhat edgy around each other, to say the least.[22]

However, Jessie and Bill soon had more problems. No sooner had they landed at Basrah than they were in turn threatened with imprisonment by the authorities for not clearing Customs at Baghdad. But, the fliers pleaded, there hadn't been time to fill out paperwork because they'd been too busy working on the plane.[23] Fortunately, they were eventually able to placate offended national egos and settle into a restful stay for a few days at the British Consulate while they waited for more engine parts to be flown out from England. Always the optimist, Jessie seized the chance to familiarize herself with the port city. At that time, the chief export from the region was dates, and consequently palms grew thickly outside the city down to the water's edge. The city's canal teeming with small sailboats reminded her of Venice, but she observed that at least half of the population was still living in primitive conditions that regularly rendered Basra prone to outbreaks of cholera, one of which had taken the lives of over a thousand people since August. Basra seemed to be acting as a central clearinghouse for the dead who, Jessie could see from her window, arrived stacked like cordwood on barges from farther upriver. The local British Medical Officer had to inspect them before they went on to the local cemetery to be buried facing Mecca. Fortunately, this particular outbreak had almost burned itself out by this time, but it would explain why the Customs authorities were so annoyed by Jessie and Bill's unauthorized arrival. In the end, they were effectively quarantined for a week there.

Although the right engine parts arrived on time, they came with the wrong fittings, for which Jessie and Bill would have to wait another eight days. Angry and impatient at the delay, they had just decided to improvise repairs and fly on regardless when news came in that there had been a raid on Nazareth by Arab bandits during which twenty-eight policemen and a number of women and children had been killed. Then one of the bombers taking off from Basrah on a retaliatory mission crashed just after takeoff, killing all on board. Shaken by the tragedies and their implications, Bill and Jessie decided to take the hint and stay where they were until the correct fittings for their parts arrived.

As the days in Basrah drifted by, Jessie and Bill became regular dinner and party guests on board the light cruiser HMS *Enterprise*, which had just been commissioned the previous year and was based in the port while serving with the 4th Cruiser Squadron. Then the Southampton flying boats caught up with them again and this time it was their turn to visit and be impressed by the amenities on board the huge planes. Inevitably, though, the correct fittings for their plane finally arrived and it was back to work. With some long hops ahead now between refueling points, they sacrificed some precious space to have an extra fuel tank fitted on the floor. Jessie would now not be able to push her stick forward enough to allow Bill to use his when taking off, and so she would have to remember to detach it each time until they were off the ground. As the *Red Rose* took off for Bushehr, some 350 miles down the Persian Gulf, they circled the *Enterprise* one last time, dropping a message to the deck that read: "We hope our enterprise will be as fine a one as yours."[24]

Flying south along the Gulf coast, they crossed the border into what was then Persia (now Iran) and followed the eastern shoreline. Jessie wasn't impressed by the barren and apparently lifeless landscape stretched out beneath her, "darkened by the shadow from the wings of death." It looked, she wrote, "as if the earth had had internal pains and the marks of her agony were left on the scored and charred surface."[25] Seeing no place at all for an emergency landing if their engine was to cut out, she grew anxious about their unreliable magneto. Yet, though the land seemed dead and grim, she could also see beautiful colored mists hanging over it that somehow intensified the hues of the yellow clay as if making amends for the desolation. Despite her fears, the engine performed well, but as the four-hour flight wore on it became rougher and noisier and Jessie was very glad to finally see Bushehr appear beneath their wings like a huge chunk of coral gleaming white on its point of land.

While now the city is dominated by large civilian and naval airports, as well as port facilities including shipbuilding and repair yards, in the late 1920s the local British administration was only just beginning to come to grips with the increasing number of aircraft landing there. Unlike shipping, which always arrived according to schedules with the correct paperwork, aircraft sometimes arrived without prior notice. Consequently, the Consul appeared a little irritated at the sudden appearance of Jessie and Bill literally out of the blue, right as he was preparing to go back to England for some leave. However, they soon got to know him and ended up staying on at his house for six days while they carried out more engine repairs. On closer inspection, they discovered that their new magneto had stripped the distributing wheel, probably due to warmer temperatures as they had flown down the Gulf. It was a case of either waiting for what could be weeks until new parts were again flown out from England or saving time by manufacturing a new wheel themselves.

They settled for Plan B, but in order to make the part, they first had to make the appropriate tools. Borrowing a local forge, and with the help of some craftsmen, Jessie and Bill quickly produced the needed tools. Then for the next five days they made test wheels from various metals until they determined from the results that a brass wheel would work the best. It wasn't quite as true as the original, but it didn't disintegrate when fitted and tested and looked like it might get them to India. So, on 3 December they waved goodbye to a relieved British Consul, who was only too glad to hand them on to the next Consul at Bandar Abbas, and took advantage of a strong following wind to keep flying south. Once again, Jessie was struck by the rugged, fierce beauty of the landscape over which they traveled. Bare mountains descended to the water's edge, shaded in pinks and purples, ranked in treeless tiers.[26] At Bandar Abbas they stayed the night at the Consulate, which was close to the airfield, and were pleasantly surprised the next morning to find that their handmade brass distributing wheel was serving remarkably well. Taking that as a good sign, they took off into the dawn for the remote port of Chabahar, near what is now the Pakistan border, reaching it in the late afternoon.

There they found a small village in which only four Europeans were living among the local population, one of whom was the manager of the local telegraph station, which also served as the accommodation and refueling point for visiting aviators. By the next morning, their following wind had become a gale blowing against them, but they still attempted a takeoff, only to snap their tail skid. Shrugging off the risk, they continued on for Karachi and its major airport, knowing that although they would have to make a careful tail-up landing on just two wheels, at least if anything broke they could get it fixed there. But the farther they flew, the more the weather deteriorated until they completely lost visibility amid wind, cloud and flying sand. Half the time Jessie and Bill couldn't even see the nose of the plane, not to mention the coast beneath them, and they became increasingly watchful for protruding mountain ranges. At intervals, Bill would carefully descend until he could see the water to make sure they were maintaining a level horizon and not flying either in circles or upside down. To Jessie it felt like they flew on interminably, becoming hungrier and lower on fuel.[27] Ever watchful of the plane's weight, they never carried extra food except for desert survival rations, and their limited fuel supply only gave them a few hours in the air at best.

Pushing the light Avro into such a strong headwind had severely cut those hours, so they were forced to attempt a landing at Pasni to refuel, only to discover as they swooped down that the airfield was underwater. There was nothing for it but to land on the steeply sloping beach, where the fliers received a warm welcome from the entire village and were able to secure enough fuel to get them to Karachi. Everyone helped to maneuver the *Red Rose* right to the water's edge to enable Jessie and Bill to make use of the flattest and firmest section of the beach. With the tip of the prop swishing the surface of the water, they took off into more bad weather. Clouds quickly banked up around them as they continued to eat sand while attempting to see the nose of the plane. Finally, after nine tough hours in the air that day, Jessie spotted the rows of flares in the distance that had been set out for them to mark the Karachi runway in the encroaching darkness. Fighting to keep the tail with its broken skid off the ground, they made it onto the ground safely. When they inspected the engine, their handmade brass wheel from Persia was entirely worn down, and the RAF mechanics estimated it would probably not have lasted another thirty minutes in those flying conditions. They wheeled the trouble-prone *Red Rose* into their hangar, where it would remain for the next few days undergoing repairs. While these were under way, the airfield was due to be visited by King Amanullah Khan of Afghanistan, who was on his way to England, from where he would embark on a European tour. A progressive monarch, King Amanullah had reigned since 1919 and had been responsible for Afghanistan's first constitution. However, as the *Red Rose* was not a military aircraft, Jessie and Bill flew it over to the Imperial Airways aerodrome, now the site of the International Airport, where the King could meet them and inspect the *Red Rose*, which by now looked, as Jessie observed succinctly, "a somewhat battered flower."[28]

While waiting for him to arrive, Jessie and Bill seized the opportunity to explore the huge airship hangar which was under construction there. When completed at a cost of nearly £4,000,000 in modern British money, the 850-foot-long, 170-foot-high and 180-foot-wide hangar would house the Imperial Airship Service's R100 and R101 airships, then being constructed in England to fly a route to the U.S. and Canada, and eventually the projected R102.[29] Jessie and Bill stood awestruck in front of this cathedral to mankind's progress in the air, craning their necks to watch the huge roof girders being lifted into position, each secured with an enormous pin that allowed enough flexibility in the structure for it to sway slightly in the wind. Then they couldn't resist climbing the flights of stairs that ascended to the bridge over the top of the front doors from where they gazed down on the *Red Rose* far below, looking from that height like a little fly with outspread wings. Clinging on to the open iron frame for dear life with the wind whistling around them, they were surprised they could be so weak in the knees when accustomed to being at 4,000 feet in a fragile biplane. Rather grateful to reach the ground again, they were just beginning to feel fed up at waiting around for the King when he suddenly appeared at the other end of the field, motoring towards them with his entourage.

Jessie was the first Western woman to meet King Amanullah Khan outside of Afghanistan; even the Queen and her ladies-in-waiting would remain in purdah until they boarded the ship for England. Her first thought was that his uniform made him look Italian before she suddenly realized that she had no idea how to greet a King, not having met one before, but he simply held out his hand to take hers and asked in French, "How do you do?" After that, it felt easy. Fascinated by aircraft, the King wanted to know if she piloted the Avro, which she was happy to confirm, and for some twenty minutes he happily examined every detail of the plane, asking a myriad of questions about the controls, engine and fittings. Then he shook hands with them both and disappeared as quickly as he'd arrived.[30] One can imagine them standing there for a minute or two after the cavalcade had roared off into the distance, while the dust settled and silence fell, then turning to look at each other as Jessie asked, "Were we just talking airplanes with the King of *Afghanistan*?"[31]

Over the next few days, a number of journalists seized the opportunity for interviews while the aviators were on the ground for more than a few hours, but it was Jessie rather than Bill who primarily attracted their attention. Headlines on their subsequent stories such as "A Woman's Flight" and "Woman's Adventures" left no doubt as to how the story of this flight was taking shape: as Jessie's story that was now being reported from Jessie's point of view as "her flight with Captain Lancaster," rather than the other way around.[32] While Bill was reported indulging in macho foolhardiness, such as swimming from the HMS *Enterprise* moored in the harbor to shore through shark-filled waters for the sake of it, readers marveled that Jessie received bouquets of roses with personal messages of congratulations from Mussolini. She did make time for her first shopping since leaving England, though, impulsively ordering a white China silk tennis dress one day and a matching white felt hat, of which she was very proud. But a while later she happened to take her hat off and leave it sitting on a box from where Bill absentmindedly grabbed it and, without looking, wrapped some of his tools in it, a typically thoughtful gesture, which needless to say did not earn him any bonus points from Jessie.[33]

Just before they were due to take off again, the Southampton flying boat group caught up with them once more. When Cave-Brown-Cave said that he would see them again in Singapore and when the officers of the HMS *Enterprise*, which had also followed them, invited Jessie and Bill to their Christmas party in Calcutta, Jessie and Bill laughed. They told them Christmas was already booked for Australia. So, at 9:30 on the morning of 14 December, Jessie and Bill lifted into the sky from Karachi in the *Red Rose*, dipping their wings in salute to the flying boats and the *Enterprise* in the bay. Although their actual flying time so far was only eighty hours, two months of calendar time had gone by since leaving England. Having accomplished half of their journey, they were looking forward to seeing Australia soon, but their return was going to be a little more complicated now.

Somewhere out there in the desert behind them, the friendship between Jessie and Bill had deepened into intimacy.[34] While today that might not sound like such a big issue, in those times of more conservative morals a relationship like this between two people who were already married could cause a scandal that would cast a shadow over their whole record-breaking venture. Worse still, Jessie was a friend of Bill's family and she'd received personal permission from Bill's wife to accompany him. Whether Kiki trusted them or whether she just didn't care, the consequences of public revelation could leave Jessie and Bill in a very unenviable position and so they became very cautious about their relationship. Jessie would always refer to Bill as Captain Lancaster and Jessie was always Mrs. Keith Miller, and for many years they always lived or stayed in separate locations. Ironically, by the time they moved into the only house in which they lived together they were no longer in a romantic relationship, or at least not from Jessie's point of view.[35] Few photographs show them making physical or even eye contact and after they left Australia in 1928 it's a rare photo that even has them in the same frame. Even more interesting, however, is the apparent tacit reciprocal agreement by the media, particularly in Australia, to not even speculate about the relationship between a celebrity couple who were constantly seen in each other's company, rather than that of their respective spouses. It's an interesting insight into an earlier age's ethics regarding privacy, compared to modern times. But Jessie and Bill had different attitudes about their roles in this relationship, and these would become more apparent and significant as time moved on.

Three

So Close, So Far Away

As Karachi faded into the distance behind them, Jessie and Bill's optimism was renewed. They felt they could face anything now. The *Red Rose* had been refitted and was flying well, and they had been able to rest and, most important, they had some money. Sir Charles Wakefield had sent them £50 and the Avro company had sent them £25 to help pay for repairs and expenses in Karachi and on to Singapore, and as a result Jessie felt much better now knowing they would not be at risk of being stranded penniless in some remote and foreign place.[1] That afternoon they reached Jodhpur, landing at the rather splendid aerodrome there to be greeted by a representative of Maharajah Umaid Singh, with whom they'd been invited to stay at his palace surrounded by miniature lakes. Once again, Jessie had to pull out her black dress that night when she and Bill were the guests of the Maharajah at a dinner in their honor, after which he was eager to show them his personal car collection of 110 vehicles, including thirteen Rolls-Royces, and to amaze them with the State jewels.[2]

The next day they flew to Agra, circling the Taj Mahal on the way in as Jessie gazed out in wonder at its breathtaking beauty. When they landed, there was no one to meet them and so, hearing there was a British regiment stationed nearby, Bill found a bike and pedaled over there to get help to tie down the plane and refuel it. The regiment was glad to offer assistance and an invitation to the Officers' Mess for dinner, after which they all walked over to see the Taj Mahal by moonlight. Jessie thought the great building was even more beautiful by night than when they had seen it from the air lit by the afternoon sun. "The solid slabs of marble gleamed like pearl in the dusky light," she wrote, "and the guides chanted so that we could hear the beautiful echoes."[3] From there they flew on to Allahabad, where they were met by huge crowds who were enthusiastic but inexperienced in their efforts to help refuel the plane. Fuel had to be pumped by hand through a hose from drums on the ground up into the tanks in the upper wing, which meant that a consistent vacuum pressure had to be maintained to achieve constant flow. These inexperienced handlers either pumped too slow and lost flow or pumped too fast and flooded the hose, causing petrol to spew out everywhere, drenching and burning Jessie, who was attempting to control the end of the hose feeding into the tank. Consequently, her skin glowed red for hours afterwards. A day later, they flew on to Patna. As they followed the Ganges at low altitude, dodging hawks and vultures, Jessie remained unimpressed. It might have been a holy stream, she wrote, but it was a filthy one down which floated hundreds of dead bodies and where crocodiles sunned themselves on the banks. "The combination of corpse and crocodile," she reflected, "is not a happy one."[4]

Since leaving England, Bill had made the most of the Avro's dual cockpits and controls to give Jessie further flying instruction at every possible moment. Consequently, her skills improved quickly and she soon became such an adept co-pilot that she often took over the controls from him. He'd simply reach over and tap her on the helmet whenever he wanted to her to take over. Jessie now had to remember that since the extra petrol tank had been fitted in Basra in front of her on the floor of the plane, this meant that she could not push her stick forward far enough to allow Bill to use his when taking off, and so she had to pull out the locking pin and remove her stick from the floor until the plane was off the ground. However, on this morning at Patna she forgot to do that. With only a short runway in front of him that ended in a huge pile of road gravel, Bill's stick jammed when he tried to pull it back and he only managed to free it just in time for their wheels to miss the hill by inches. Rising into the sky on their way to Calcutta (now Kolkata), they dipped a wing to the huge crowd gathered below who were entirely unaware they had very nearly witnessed an abrupt end to a pioneering journey. Then, as they began their descent into Calcutta, Jessie had more joystick trouble. When she took out the locking pin, the stick stayed firmly wedged in place! As Bill climbed again, Jessie kept struggling but only succeeded in jamming the stick between the fuel tank and the bottom of the cockpit instrument panel, which put the plane into a steep nosedive. Jessie was frantically pulling and twisting on the stick, while Bill was banging her on the head screaming to her to pull the stick out because it had also disabled his own controls. To Jessie's horror, the *Red Rose* was diving so steeply that when she looked up she was staring over the top of the fuel tank in front of her at a rapidly widening patch of ground that seemed to have her name on it. With both feet braced on the floor, she wrapped both hands around the stick and tried frantically to move it. In desperation Bill turned off the engine, but still the plane fell out of the sky like a stone at an alarming speed. Jessie was furious with herself at not being able to pry the stick loose until, only a few feet from the ground with nothing to lose, she kicked the stick sideways with both feet and it came out. She would have done anything, she wrote later, rather than slam into the ground still screaming that her stick wouldn't move.[5]

Whipping the leaves from trees, they leveled off over a ridiculously peaceful Indian landscape that looked like a patchwork rug made out of colored pocket handkerchiefs. "If green is the fashionable color, then India is surely in the forefront of styles for her dress is in every imaginable shade of that unlucky tone," Jessie wrote, revealing another of her superstitions.[6] To her, those little fields seemed be continuously at risk, because so much of the land was being covered in marble that the ordinary people were being left with less space in which to grow their food. As they flew over the outskirts of the city, at that time the great jute market of the world, the quilt of fields gradually disappeared under layers of polluting smoke from factories, but jutting through the haze were the towers of mosques that seemed strangely out of place beside the bustle of a modern industrialized city. They set down at Dum Dum Airport (now Nataji Subhash Chandra Bose International Airport), a few miles from the center of the city, where they were met by Bernard Leete holding a bottle of cold beer in each hand. In January, Leete and T. Neville Stack had been the first pilots to successfully fly light planes from England to India, for which they were both awarded the Air Force Cross, and Leete had remained here, operating airmail and charter services and establishing flying clubs. Jumping out of the plane, Jessie and Bill toasted solid ground under their feet again and that Jessie had just broken two world records: she had now achieved the longest distance flown by any woman in any type of airplane anywhere and the longest distance in the air by a woman with someone else.[7]

After a brief rest, Jessie and Bill went to work on refueling and engine maintenance to ready the *Red Rose* for a morning takeoff. Throughout the trip, Jessie worked alongside Bill on the plane as a mechanic as well as in the plane as a co-pilot. Standing on the fuselage, she would pour petrol into the wing tanks from cans, straining it through chamois leather to ensure it was free from dirt and water. The wind would inevitably blow the petrol back over her, leaving her suffering from burns on her exposed skin much of the time. She would clean the spark plugs and filters, reoil engine parts, chip the carbon from the cylinders, and before takeoffs she would walk around the plane to visually check it and then walk the landing strip to look for any obstacles or potholes. When they finished their working on the Avro, Jessie and Bill gratefully climbed into Leete's car and drove to their hotel through traffic that moved at the pace of the lumbering water buffaloes everyone used as transport. After they had settled in, Leete took them to meet the manager of a touring Australian circus. "Anything which had even smelled gum leaves in passing was sure of a welcome from me," Jessie joyfully declared.[8] They all had a wonderfully entertaining time at that night's performance, after which the manager invited them back to his tent for supper, where he demonstrated his favorite trick: tossing a series of tiny boomerangs one after the other across the back of his neck, he could keep them circling for some minutes above his head like a halo. He handed some to Jessie and tried to teach her, but no matter how much she tried, she couldn't learn the knack. Early the next morning, a small crowd gathered at the airport to see them off. While Bill was warming up the engine, Leete mentioned to Jessie that while she'd had to pack her bag and the plane herself, Bill had been attended by "a whole tribe of people" packing his bags and putting on his shoes. "No doubt," she recalled tartly, "that was why he didn't do any thinking for himself."[9]

They finally left Calcutta, flying out over the Ganges delta, a nasty bit of country that was mostly thick swamp. Jessie prayed that the engine wouldn't falter; no one would ever find them if they went down here. Then it was over the Bay of Bengal to Akyab, now Sittwe, Myanmar. On the ground, Jessie discovered why Bill should have been thinking for himself back in Calcutta. When he went to pay for their fuel, he discovered there was no roll of pound notes in his pocket. He turned to Jessie and asked: "Have you got it?"

Jessie took a step back, put her hands on her hips and gave him a very long look.

"No, Bill," she said tightly. "I don't have it. If you'll recall, I wanted to have it but you told me I'd only lose it. 'Let me keep it under my pillow for safekeeping because I'm such a light sleeper,' you said."

Bill had the grace to look sheepish and the sense to keep silent.

Now faced with one of her worst fears, one that had constantly haunted her, Jessie understandably lost her temper.

"And that's where it still is, Bill, isn't it?" she yelled. "Under your bloody pillow! If you hadn't grown such a swelled head back in Calcutta and needed servants to tie your own shoelaces, you might have remembered it! What are we supposed to do without any money?"[10]

It was a very good question: in a time before credit or debit cards, if you didn't have cash in your hand or a personal letter of credit to a bank manager in a foreign country in 1927, you had nothing. Not only that, but with the money was their airplane mascot: a small Saint Christopher medallion given to them on their departure by Bill's mother. Fliers are notoriously superstitious when it comes to lucky charms, and so the loss of this medallion was serious. The only thing they could do was telegraph Leete back in Calcutta, which is what Bill promptly set off to do with the last ten rupees they owned and Jessie's angry words ringing loudly in his reddened ears. He sent cables to Leete and the Police Commissioner, who promptly checked with the hotel, but it was all too late; the money and the medallion had gone. Luckily, Leete could vouch credit for them and they could fill their fuel tanks for the leg south to Rangoon, now Yangon.

On the way to Rangoon, they flew over some of the most beautiful country Jessie had seen. "The sea was a magnificent blue," she wrote, "dotted here and there with lovely little green islands. Stretches of white, curving beach were an invitation in themselves to come and sun-bake on the sand."[11] Beyond the beaches and mangrove swamps were valleys of irrigated paddy fields between the mountains, and then the Irrawaddy River was twisting its way through heavily wooded country below them. As they began to climb over mountains 8,000 feet high, thermals and updrafts buffeted the *Red Rose* about the sky for some time. On the other side, they flew in towards the city over more paddy fields. Jessie knew only two things about Rangoon: that they had to land at the racecourse and that dominating the skyline of the city was a wonderful gilded stupa known as the Shwedagon Pagoda, reputedly the oldest Buddhist stupa in the world.[12] They were down to about 2,000 feet and within sight of the sun glinting on the peak of the stupa, and Jessie was just thinking that places like this must be centers of peace and repose, when the engine began to bang loudly and then abruptly stopped. Once again, they started to drop from the sky in silence. "When I get back to England," Jessie said to herself, propping her feet up off the floor so her ankles wouldn't be broken if they crashed, "I really must take up the Avro's limited gliding distance with the designers."[13] A cylinder or piston must have broken, she thought, and so they were obviously not going to restart the engine. The plane wouldn't stay in the air long enough to make the racecourse, and so they would have to come down in a paddy field. If they could find a field with enough length, the soft plants, soil and water should act as a brake on their landing speed before the plane was too severely damaged and, fortunately, that is what happened. Even without their Saint Christopher medal, luck had evidently remained with them this far. Just grazing one of the mud embankments between the fields, they came to a stop in a shower of mud, water and rice plants. Hardly able to believe they were still in one piece, they clambered out and checked themselves and the plucky *Red Rose*. They and the plane, it seemed, had survived intact.

It was about three-thirty in the afternoon and unbearably hot. To their amazement, no one seemed to have noticed that an airplane had just fallen out of the sky into their fields. When Jessie and Bill climbed up on an embankment and looked around, everyone was still unconcernedly cutting or planting rice. It was only Bill's waving and shouting that suddenly prompted a rice-farmer stampede from every direction and the next minute they were frantically trying to keep people from climbing on the plane. It was all too easy to put a foot through the fabric if you didn't know just where to step, and they were only carrying a limited supply for repairs. Finally, they found a Burmese who could speak English and Bill walked with him to the nearest railroad station and a telephone while, once again, Jessie guarded their plane. For some two and a half hours in that scorching heat, with steam rising from the marshy ground of the paddy fields, Jessie chased curious Burmese round and round the machine, pushing off the ones who wanted to see how it worked from the inside. It was a hopeless task. While she was on one side, a flank attack was made on the other. Fortunately, they weren't trying to remove any parts: because most of them had never seen a plane before, they were just trying to discover what it was made of by pushing their fingers through the fabric. Although the official who had gone with Bill had left two of his men behind to help protect the plane, they just thought Jessie's efforts were all a huge joke. Soaked by perspiration, parched with thirst and becoming angrier with the world by the minute, Jessie was thoughtfully eyeing her pistol when an English couple by the name of Tait arrived, just in the nick of time, carrying baskets of iced soda water. They lived some distance away on the other side of the river but had set out as soon as they heard of the aviators' plight, and now they placed their own guard on the plane until a group of police came out from Rangoon to take charge.

When Bill returned, he and Jessie walked back through the rice fields with the Taits to their home. Meanwhile, the news had been telephoned to Rangoon of the forced landing, but somewhere along the telegraph lines the news became much more dramatic than reality. When they reached the Tait house, the servants were already fielding calls from journalists prompted by an international Reuters News Agency report that the crash had killed them both! Knowing their friends and relatives in England and Australia would be devastated at hearing news of a fatal accident, they rushed to the telegraph office to send word that all was well and they were still very much alive. Jessie then stayed on with the Taits while Bill was accommodated with the air survey crews at Monkey Point, a few miles downstream at the junction of the Rangoon (Yangon) and Bago rivers. This junction is more than just a meeting place of rivers: it's also the location of the sunken Great Bell of Dhammazedi, a story that fascinated Jessie. Reputedly the largest bell casting in history, of a mixture of silver, gold, copper and tin that weighed around 290 tons, the Great Bell was cast in 1484 at the order of King Dhammazedi as a gift to the Shwedagon Pagoda. Looted from the Pagoda in 1608 by Portuguese warlord Filipe de Brito, the bell was being floated down the Bago River when the raft broke up when it reached the junction. The bell sank into the mud of the river bottom, never to be seen again. Although many have tried to locate it, no one has yet been able to successfully salvage the Great Bell.

Jessie and Bill worked in the sun all day on their plane, discovering that the piston rod had shattered and so they were going to need a new cylinder as well. They sent another frantic cable to Leete, who immediately sent a secondhand piston and cylinder from Calcutta. Meanwhile, unable to get the loss of the Saint Christopher medal out of her mind, Jessie kept up her spirits by writing optimistic cables to Australia that they "still had their tails up." Optimism wasn't money, though, and that was what they needed right now.

Too proud to let anyone in Rangoon know they were flat broke, Jessie finally had to break down and cable her husband, Keith, back in Melbourne. She kept it short and to the point: "Miller, Herald, Melbourne. Cashless."[14]

Perhaps they looked like they needed some cheering up, because the very next day they were invited to the races by Mr. Edwards, the chief executive of Steel Brothers, rather ironically at the very track where they had failed to land the *Red Rose*. Jessie managed to borrow a suitable outfit and off they went. While she didn't back a winning horse, Jessie did find a winner of another kind. She met Mr. Ebenezer Millar, who for reasons of his own, despite having never met them before, became their financial benefactor. With Jessie living on one side of the river and Bill on the other without a rupee between them, Mr. Millar was happy to provide a launch and a car for them while Jessie waited in vain for money or even a word from Keith. The Christmas holidays arrived, but there was no gift-giving for the two aviators. For two days, they went without food. Of course, people would ask them out, but Jessie and Bill did not like to keep on accepting invitations that they could not return. On one or two occasions, they even dashed down to Monkey Point and bought up bananas from the air force mess. As soon as the holidays were over, Jessie asked Mr. Tait to call the bank and see if there was any money there. The reply came over the wire that there was £30 in their account. It had been there right through the holidays while they hadn't been able to access it. On New Year's Eve, Professor Egger, the government advocate, gave a dinner for the flying duo. A feature of the decorations was a model of the *Red Rose*, which he had made himself, electrically operated and complete with a tiny, buzzing propeller. As the party was fancy dress, Jessie decided to make fun of their situation. She cut the legs off a pair of white pants and frayed the edges, borrowed a torn white tennis shirt, knotted a piece of rope round her waist and attached a bottle to it. Donning a battered panama hat, she poked her toes out of some decaying tennis shoes and tied a red kerchief around her neck to complete the image of a disreputable-looking old beachcomber. She swore it was a New Year's Eve she would never forget.[15]

The piston and cylinder duly arrived, and with the aid of the air survey ground engineer at Monkey Point, Mr. Bishop, the engine was quickly reassembled and they were ready to put it back into the plane, but the *Red Rose* was still sitting out in the middle of the paddy field. Getting it out of there wasn't going to be easy. Jessie and Bill calculated that with the plane completely emptied of anything not bolted into it, they'd have just enough petrol to fly the plane out of the field to the racecourse. The pragmatic air survey crew offered to winch the plane onto the back of one of their flatbed trucks and transport it to the racecourse, but of course proud pilot Bill wouldn't hear of that. No airplane of his was going reach its destination on the back of a truck; he was going to fly it there or wreck the plane trying. To minimize weight, Jessie remained on the ground. Her job would be to mark the position of the embankment by standing on it, indicating for Bill the end of his runway and the height he'd need to be at to clear the bank. Usually, the light Avro could take off within quite a short distance, but this was a wet, muddy paddy field, not a dry, hard airstrip. Consequently, Bill's speed would be slower and so he'd need a lot more distance to build up enough speed to get off the ground. Standing up there on the embankment watching Bill jockey the plane into position, Jessie realized that she was more scared than she'd ever been in her life. Had she been in the plane, she would have been too busy to think about it, but just standing here facing an oncoming plane rushing towards her was truly terrifying. For a small plane, it seemed to get very big before the wet ground reluctantly released its grip on the wheels and Bill sailed over her head towards the racecourse, where he landed successfully and they could refuel.

Before they left for Tavoy, their friends and well-wishers gathered around to say good-bye. Mrs. Tait took Jessie aside and gave her some important advice: "Don't forget, before you leave have a good look inside the plane for snakes. Your machine has been sitting out in the middle of that field for some time, and all these swamps are infested with snakes. You might find that you are carrying an extra passenger if you do not make a search. We have highly poisonous kraits here and a bite from one of those can kill you within a few hours if you don't get attention." Jessie naïvely thought she was probably just trying to scare the tourists. Thankful to have the engine running again and money in her pockets, she then forgot Mrs. Tait's words amid the bustle of preparation. As Rangoon fell away under their wings, Jessie and Bill became intent on plotting their direction to Tavoy. Suddenly, about ninety minutes into the flight and well out over thick jungle, Jessie felt the plane start to dive deeply and then lurch all over the place. Quickly turning around, she screamed back to Bill: "What's the matter?" Pointing violently down at the floor, he yelled back one dreaded word: "Snake!"[16]

Suddenly, Jessie recalled Mrs. Tait's words. She hadn't been joking after all, and Jessie realized with a guilty pang that she should have checked the inside of the plane as well as the outside. Now they had a very dangerous stowaway. Bill quickly took his feet off the rudder control in his efforts to dodge the reptile and stamp on it, but the snake merely slipped by him and wriggled through into Jessie's cockpit. Jerking her feet off the floor in turn, she saw it was unmistakably a deadly krait with distinctive white bands, about three feet long. Without a second thought, she whipped out the dual control stick and cracked it over the head hard, beating it another two or three times just to make sure it wasn't going to revive. Blood spurted all over the cockpit as, with shaking hands, she flipped the writhing body up off the floor and threw it over the side of the cockpit out into space. Leaning over the side, she watched the krait disappear towards the green canopy of jungle below them. The incident would become part of Jessie's legend; she became famous as the lady who fought a deadly snake in a biplane cockpit with her bare hands and won. The story became a public relations media triumph, the original "snake on a plane" of the 1920s, announced in headlines such as "Aviation Snake in Plane," "A Flying Snake," and the author's personal favorite, "Snake Goes Joy Riding."[17] Even the London *Times* devoted a half column to the incident, commenting that if a reptile had really hoped it could help mankind appreciate snakes as a goodwill flier, then it should have arranged more preliminary publicity.[18] Jessie would be forever remembered as the aviator who actually killed a snake in mid-air with, naturally, her joystick.

When they had been in Karachi, the fliers had consulted about their route with experienced pilots there, and the general consensus had been that there was nowhere to land and refuel between Rangoon and Victoria Point (Kawthaung) on the Thai border. As it was a 650-mile run and the Avro Avian couldn't carry that much fuel, they would have to fly to Maungmagon Beach, twenty miles outside of Tavoy, land on the wide beach at low tide and refuel from drums that could be stored there. High tide would be at about three o'clock, according to the experts, so they had left Rangoon in time to arrive as the tide was going out. Unfortunately, the only chart that could be found in Rangoon from which to make the tide calculations was dated 1882; consequently, when they flew over Maungmagon Beach they could clearly see that the tide was at the full rather than being out. Too hungry and tired to wait for a tide, they continued flying about twelve miles farther along the coast until they found another beach on which they could land. As they coasted to a stop with one wheel in the water, the other on the sand, hundreds of Burmese emerged from the trees and, sitting down, quickly encircled them in a wide ring. There was not a European face to be seen anywhere in the vast crowd, who, talking loudly with one another and pointing upwards, seemed to think the *Red Rose* was some strange bird fallen from out of the skies. Fiercely hungry by now, Jessie and Bill managed to convey their need for food, whereupon the local people presented them with opened coconuts and turtle eggs. Jessie of course had never seen turtle eggs before and wasn't entirely sure she could keep them down, so when their hosts weren't looking she and Bill quickly buried the eggs in the sand, preferring to eat some bananas instead.

It seemed a good time to take stock of their position, as they were now in something of a quandary. Jessie and Bill could not leave the plane unguarded, as the fabric had already been torn in several places and the plane would not be flyable if any further damage was down by curious spectators. However, Maungmagon Beach was twelve miles back along the coast and Bill didn't want Jessie to walk that distance at night, but nor did he want to leave her to guard the plane alone among all these unknown native people. Eventually, having talked it over, they decided to spend the night on the beach together with the plane and wait for the tide to turn at ten the next morning, when there would be enough light to take off. It was a bitterly cold nightmare of an evening. They were bitten by sand flies and chased by land crabs while taking it in turns to guard the plane. They decided against swimming in case there were sharks in the water and so amused each other flashing out S.O.S. messages into the darkness with a powerful torch. As they grew hungrier by the hour, the night stretched out into misery. They couldn't even get any sleep, surrounded as they were by hundreds of natives who chattered among themselves incessantly. Finally, dawn's light revealed there was enough exposed sand for them to take off. As the beach fell away beneath the *Red Rose*, Jessie and Bill could see the crowd still gathered there, still looking up into the sky. From there, it was only a short hop to Maungmagon Beach, where an increasingly anxious group of Europeans who had waited for them all night were happy to see them touching down. Jessie and Bill were more concerned about eating. They sent cables to Tavoy for fuel, but it couldn't arrive until three in the afternoon, and so they settled in for some serious attention to food. When their fuel finally arrived, Jessie supervised the refueling operation as usual. However, Victoria Point was at least five hours' flying time away and darkness would have long fallen by then, so they accepted an offer to stay at the home of an air force officer and his wife until the following day.

Taking off from the beach next morning, they climbed above a dense cloud bank at 4,000 feet to find blue sky and sunshine. "The clouds looked so solid," Jessie wrote, "that one could almost have landed.... As the sun caught the propeller, a halo seemed to form around the machine and was reflected against the clouds."[19] Landing at Victoria Point, they filled the *Red Rose* to the absolute limit of the tanks. In these hot conditions, though, they would need about a 400-yard run to get the fully laden plane off the ground, and the strip at Victoria Point was quite short, with a drop at one end and a large hill topped by trees at the other. Despite some doubts, they struggled into the air safely and were just beginning to congratulate themselves when they realized Bill had packed away the maps needed for the route to Taiping. Totally ignorant of the local geography, they had no choice but to land again and find the maps. Once again, Jessie reminded Bill what she thought of his organizational capability. After all, as he was the pilot it was his responsibility to select the map he would need from their collection and make sure it was put in the cockpit while Jessie was checking the fuel tanks and carrying out her walk-around inspection. Neither of them spoke to each other all the way south into the Malacca Strait to Penang.

As they circled the town's racecourse, it quickly became apparent that if by some miracle they did manage to land on it, they'd never be able to take off again because the track was too closely surrounded by trees and houses. Even the beach was too cluttered with boats and nets to risk touching down there, so reluctantly they had to fly on to Taiping, the capital city of Perak in what was then known as the Federated Malay States, thankful they had taken enough petrol on board. There, on 5 January 1928, they were able to land on the polo ground, despite torrential rain, and refuel and work on the engine in preparation for an early-morning takeoff. However, when Jessie inspected the plane by the light of dawn the following morning, she noticed the top petrol tank had sprung a leak. All their work the day before was wasted. The tank would have to be drained and removed from the plane to be repaired, so they'd not be going anywhere anytime soon. Instead, Jessie and Bill had to set to work draining the tank into tins and then lifting it out, becoming soaked in fuel as they did so. Having cleaned it, they set to work soldering the hole in the searing heat while trying not to ignite themselves in the process. They looked a piratical pair as they worked away: an unshaven Bill was dressed in shorts and a filthy shirt fastened with a large safety pin with his bare feet stuffed into tennis shoes while Jessie, likewise in shirt and shorts, was barely visible beneath a coating of grease and oil streaked by perspiration. With the leak repaired, they bolted the tank back onto the plane, refilled it, repacked the plane and were finally ready to go. By now some four hours late, a dog-tired Jessie and Bill just climbed into the plane and took off without even bothering to clean up. Besides, they'd been told that there were not enough Europeans at Kuala Lumpur to worry about dressing properly.

Afterwards, they were never quite sure whether the people at Taiping took advantage of their lack of local knowledge and played a joke on them or whether they were just that contemptuous of Kuala Lumpur society, but when Jessie and Bill landed there at noon, after flying over tin dredges and rubber plantations hacked out of the jungle, the two fliers received a surprise. It was, of course, the seat of government and although the governor, Sir Hugh Clifford, was away at the time, practically the whole staff of Government House turned out to meet them, including the Chief Secretary, the Hon. H.W. Thomson. The entire aerodrome was packed with beautifully dressed people there to meet a fuel-soaked and grease-covered Jessie and Bill climbing out of their plane looking like a couple of old tramps. As they were being officially welcomed and introduced to the officials present and invited to be guests of honor at Government House, Jessie just wanted to crawl back into the plane and get out into the jungle. "I have never felt so terrible in all my life," she remembered later. However, everyone seemed to understand the aviators' position. When they arrived at Government House, Lady Clifford came to Jessie's room and asked if she had any clothes, prompting Jessie to confess that this time Bill had wrapped his spare engine parts in her one and only tennis frock and she didn't think her black chiffon dress was the thing for luncheon wear in the tropics. A suitable frock for lunch was promptly borrowed from a teenage guest, because Jessie was so small, and the black chiffon was rescued and laundered for dinner that night. As she sat there playing bridge after dinner, Jessie felt like she was back in civilization at last.[20]

Rested, scrubbed and dressed in clean clothes again, Jessie and Bill left for Singapore the next morning, Saturday, 7 January. Two and a half hours later, in deteriorating weather, they landed there at around ten-thirty in the morning in rain so heavy that it stalled the engine as they taxied to a halt. At that time, this wide expanse of flat, open ground was often used as a landing field but was prone to becoming waterlogged in heavy rain. As they taxied in, a huge crowd converged on the tiny plane, churning the wet ground into a quagmire. By now well over halfway through their journey, having covered some 10,500 miles, the two aviators had established a new long-distance record for a light airplane. The *Red Rose* was the first aircraft of its size and type to reach Singapore from England, and Jessie was the first woman to fly to the Far East. As the police held people back, Jessie and Bill were welcomed by the official party: the Colonial Secretary, Sir Hayes Marriott, who was Acting Governor at the time; Major Lowe, representing the Commanding General; Mr. W.F.M. Churchill, private secretary to the Governor; Mr. R. Johnson and Mr. H. Chapman, captain and secretary of the Singapore Flying Club. As their shoes filled with water while they stood there, Bill tersely informed the official party that it was a dreadful landing ground, especially for a place the size of Singapore, and he went further by telling the *Straits Times* journalist that in fact it was one of the worst landing grounds they had encountered so far. Worried they might not be able to take off if the weather deteriorated, he and Jessie discussed with the officials moving the plane to the nearby Serangoon Road racecourse, just to the southwest of them, but the Turf Club assured them the track was just as prone to flooding as Belastier Plain, so they tied the *Red Rose* down for the night and hoped for the best.[21]

Jessie, dressed in her tropical flying outfit of khaki shirt, shorts and topee, was already becoming tired of the inane questions of journalists who persisted in seeing a woman rather than an aviator. However, she gritted her teeth and admitted that she was looking forward to returning to her home in Melbourne, that she did regularly fly the plane and that she was even quite capable of refueling the plane and dealing with some of the mechanical problems. In fact, it was because of engine mechanics that Jessie and Bill decided to stop over in Singapore for the weekend. Always looking ahead to the long leg they would have to eventually fly over the open water of the Timor Sea to Darwin, they decided to use Sunday to carry out some much-needed maintenance and then leave Monday morning. The Singapore Flying Club had arranged a luncheon at Raffles Hotel and Sir Hayes and Lady Marriott invited them to dinner that night and to stay at Government House, so they had little time to themselves the rest of the day. Jessie had to rescue her white felt hat from where it had been wrapped around Bill's tools, push it back into shape and clean it of most of the oil spots. Then she sent off a wire to Sir Charles Wakefield reassuring him the engine was running just fine on the fuel and oil his Castrol company was supplying. Most of Sunday was spent overhauling the engine and closely inspecting the *Red Rose*. From here on it would be hard flying, and a lot of it over the open sea, so they wanted to make sure everything was going to work just the way it should. Bill had come up with an emergency plan to carry some thirty car tire inner tubes that he could rapidly inflate with a quick action pump and then lash together with rope to the plane to keep it afloat if they had to ditch at sea. He would leave his exact course details in Singapore, along with his cruising speed and estimated time of arrival, so that if they were more than four hours late arriving in Darwin then ships could be readily notified to search along that path and the aviators would be carrying a powerful torch in order to flash signals if necessary.[22]

On Monday morning, 9 January 1928, Jessie and Bill took off from Singapore and circled over the harbor before turning the plane's nose south across the Singapore Strait and the South China Sea towards Muntok. Looking down at the west coast of Sumatra, Jessie realized they must be crossing the equator. In fact, Jessie was the first woman to cross the equator in the air.[23] Beneath her, the dense jungle "seemed a flattened green rug. The trees merged into one another so thickly that the landscape seemed to have compressed breadth but no depth."[24] Beneath them, palm-fringed bays alternated with mangrove swamps, and tiny emerald islands with a lacy edge of white dotted an absolutely empty coastline. The only sign of animal life in this entire tropical landscape was a herd of elephants Jessie briefly spied through a gap in the trees. Eventually they reached Bangka Island and a ghastly aerodrome at the port of Muntok on the northwestern tip of the island, where the grass on the sloping airfield was nearly up to Jessie's waist, completely obscuring any holes or hillocks.[25] The Dutch people there turned out in force to meet them and their hospitality knew no bounds; at least a hundred people attended a reception that night for Jessie and Bill. Unfortunately, although a new radio set had arrived for the occasion, it couldn't be made to work despite much effort and so they weren't able to contact anyone to communicate their whereabouts.

As usual, Jessie carried out her morning inspection before they took off into the cool of early morning, waving farewell to the whole community. Suddenly, at about 150 feet in the air, the engine cut out completely. For a moment, Jessie thought Bill had forgotten something again and was turning back to land, but the engine didn't start. With a sickening sideways lurch, the plane began to fall out of the sky like a leaf falling from a tall tree. With no power, they had no control; the only thing Bill could do to try and save them, especially Jessie seated in the front cockpit with one fuel tank above her head and another at her feet, was keep one wing down so it would act like a shock-absorbing crumple zone when they sideslipped into the ground. For Jessie it felt like hours went by until they smashed into the ground with a grinding crash and somersaulted down the slope. The undercarriage ripped off as the plane turned over and then the world went very still.

Jessie shook her head to clear it from the momentary shock and discovered she was hanging upside down, imprisoned in a web of tangled wires and wreckage and drenched in petrol from the ruptured tank that was now above her. Hearing the petrol sizzling as it hit the hot engine, she imagined an imminent flaming explosion and began a desperate struggle to extricate herself. When she found enough breath in between hammering and pulling, she called out to Bill, but there was no response. Unable to see around her, Jessie could only conclude he'd been crushed in his cockpit. With her arms pinned by wires, blinded by petrol and blood, she frantically pushed down with her feet and out with her hands until she was able to create a small gap in the fuselage and then, taking off her helmet, she wriggled down through the hole. Much like a screw being twisted out of a plank, she popped out through the wreck feet first onto the ground, bleeding from innumerable scratches and abrasions but happy and amazed to be still alive.

Quickly pulling herself together, Jessie lurched to her feet and started frantically searching through the wreckage for Bill, but he wasn't in it. Figuring he must have been thrown out somewhere, she then started to walk around the crash site in a gradually widening circle until she eventually found him about twenty feet away lying unconscious, facedown in the grass. Turning him over, she was horrified to see blood pouring from his mouth. Thoughts of dreadful internal injury went racing through her mind as she propped him up so he wouldn't choke. Finally, his eyes flickered as he began to come around and she knew that at least he wasn't dead, but then he passed out again. Jessie found some water and began to wash the blood off his face, discovering as she did so that his extensive bleeding was because his teeth had gone right through his lips. She gently pushed them back out again, but Bill continued to bleed heavily. Just as she was slapping him into full consciousness, the Dutch settlers finally found them and, despite Bill's protests that he wanted to see his plane, they loaded them both in a car and rushed them to a hospital. There she learned that her eyes had been blackened and her nose had been broken; worried about Bill and pumped with adrenaline, Jessie hadn't felt a thing. "Only 1800 miles from Australia, and now we have smashed our bus," she lamented.[26]

Stranded in Muntok with a smashed plane, Jessie and Bill were more concerned about the *Red Rose* than themselves. Once a doctor had stitched Bill's cut lips and plugged and patched Jessie's nose so it would retain its shape as it mended, they were off to the crash site to take photographs and assess the damage. It was severe, so severe that "all the heart seemed to go out of us as we totalled up the damage," Jessie recalled later.[27] Still, she was committed to an adventure from which she was not going to be deterred by a small thing like a plane wreck. Airplanes were just machines, after all, that could be fixed or replaced and she was soon firing off telegrams for help so they would be able to complete their journey. She wired to Avro asking for a replacement plane and also telegraphed newspapers in Australia to place advertisements asking if some sponsor would be prepared to back them with £700 for a new plane. Neither bid was successful. Not a single Australian offered them even one pound. In the end, all they could do was dismantle the remaining sections of the *Red Rose* and pack them up to await shipment back to Singapore. Jessie cabled the Acting Governor, Sir Hayes Marriott, and his wife in Singapore: "Heart broken, broken aeroplane." They replied: "Come back immediately. We will meet you."[28]

So, the two bruised and bandaged fliers came back in torrential rain to Singapore, this time aboard a local Chinese junk, where the Marriotts were waiting at the dock to bundle them into their car for the drive to Government House, where Jessie and Bill stayed to recuperate. There they received a longer letter from the Avro company, which reluctantly explained that, after some consideration, they'd concluded it simply wouldn't be good public relations to send another plane. It might look as though they were admitting there was some kind of fault in the Avian's design. For their venture to be really successful advertising, Jessie and Bill had to complete the journey in the same plane in which they had set out in order to promote the Avian's durability, so it was back to the workshop amid a flurry of press predictions that their flight was probably over.

While the repairs were going on, although it was more like a rebuild, Jessie moved in with an Australian rubber broker, Keith Bond, and his wife while Bill went to live at the new Royal Air Force base under construction at Seletar. Ultimately, they would be in Singapore for three, long, frustrating months that became even harder to bear when they heard the news that "Hustling" Bert Hinkler had left England in an identical Avro Avian to claim the solo flying record from England to Australia. It would mean that, after all their efforts, they would not be the first to complete the entire distance between England and Australia in a small aircraft. Even though Bill kept a stiff upper lip about it all, Jessie was quietly furious. As far as she was concerned, Hinkler made the attempt because he had heard of their accident and knew they'd be unlikely to get to Australia before him if he left England immediately.

While Hinkler's timing for his flight might look opportune, the situation was not really that black and white. Born in 1892 in Bundaberg, Queensland, Hinkler had been fascinated by flight from an early age, building and flying gliders and early airplane prototypes. By 1914, he was working for the Sopwith Aviation Company in England, before joining the Royal Naval Air Service during World War I. After the war, he was employed by the A V Roe aircraft company as a mechanic and test pilot and in the middle of 1920 he attempted a flight to Australia in a "Baby" Avro that had a wingspan of only 25 feet and a 35hp engine with a cruising speed of 65–70 miles per hour. Breaking the record for a non-stop flight in a light aircraft as he flew from London to Turin, Hinkler got as far as Rome only to find that conflicts had broken out in Syria and Egypt that would delay him flying across those areas indefinitely and so he returned to England. The following year, Hinkler and his Baby Avro finally completed the journey to Australia by ship, where he created another long-distance non-stop flight record from Sydney to his hometown of Bundaberg. Then Hinkler returned to Avro as their chief test pilot, contributing to the design of their new Avro Avian. However, he never forgot his ambition to fly to Australia. In 1927, Hinkler broke his own non-stop record by flying to Riga in Latvia and, having proved that the Avro Avian was capable of flying anywhere, the methodical and experienced Hinkler began making some design modifications to his Avian 581 prototype, that would improve his chances of reaching Australia quickly and in one piece. He widened the undercarriage for stability and to improve access to the motor, and he added a second, larger petrol tank with a hand pump to enable fuel transfer in flight, giving him a capacity of 66 gallons and the ability to stay aloft for sixteen hours if necessary. Hinkler also modified the tail skid so it could be steered and the propeller so it could be spun single-handed to start the plane.

Hinkler didn't leave England until 7 February, arriving in Darwin on the twenty-second, so he certainly had time to hear the news of Jessie and Bill's accident, make his decision and organize his flight. That he worked for Avro in such a significant position might have also worked in his favor; he might have pointed out to management that if Jessie and Bill didn't make it, then Avro would lose their chance for publicity of the Avian. Flying the same type of aircraft as Jessie and Bill, although a much earlier version, Hinkler had more proven experience as a long-distance pilot and a better plan, and so he would have looked to Avro like their man for the job. Nevertheless, whatever he might have thought at the time of Hinkler's opportunistic leap into the cockpit, Bill Lancaster was a man of honor who would hear no wrong said of a fellow "knight of the air," not to mention that Hinkler's wife, Nancy, and Kiki Lancaster were also the best of friends. When Hinkler arrived in Singapore on 19 February, only twelve days after he'd left England, Jessie and

Bill were there to graciously meet him and to be his hosts for dinner that evening. When Bill heard that some angry Australians were planning to sabotage Hinkler's plane so he'd be delayed until Jessie and Bill could resume their flight, Bill gallantly sat guard in the Avian overnight to ensure its safety.[29]

When the crates of *Red Rose* pieces had arrived from Muntok, the Harbour Board kindly loaned Jessie and Bill the use of a shed along with some Chinese mechanics. After some careful examination and inventory of missing parts, Jessie and Bill concluded that the fuselage was in better shape than it looked; reassembly would be a viable option. The mechanics concluded that the accident had occurred because good old absentminded Bill forgot to turn on the valve from their main fuel tank, although the cause given out to the press was that the plunger axle broke. So, the engine had run only as long as the fuel in the carburetor had lasted, and then it stopped. Having cabled to Avro in England for the parts they needed, they had to patiently sit around in Singapore until the spares arrived. With their money pretty much gone now, "sitting around" accurately described their days. For a week they existed on a dollar between them, but they kept their plight a secret out of pride. Then Keith Bond, who worked for Fraser & Co., the oldest share-broking company in the Near East, happened to mention Jessie and Bill's troubles to his boss, the wealthy Sam Hayes, with whom he quickly set up a meeting. Hayes was so impressed with the pair that he offered to back them financially for the rest of the trip, including their repair bills in Singapore. In return they vowed to pay him back every penny once they got to Australia, where any number of sponsors would hand over money when they landed. After that, their situation and their social life improved. Jessie's black evening frock was put back in use and, tired of wearing the same shirt and shorts outfit, she went out and bought some cotton dresses.

After a wait of six weeks, the spares finally arrived and Jessie, Bill and their mechanics set to work in earnest. Eventually the repair work moved out to Seletar under the supervision of Squadron Leader Freeman and his men. As in any rebuild, the new parts didn't quite fit the old, no matter what the work manuals said, and had to be reshaped and rewelded as the work dragged on day after day. While they were working, they heard news on 28 February that the Supermarine Southampton flying boat group with whom they had started out were about to be the first planes to arrive at the Seletar base. Dashing down there, Jessie and Bill were the first to greet the officers as they landed. It was a happy reunion of a unique group of people who at that time were engaged in the longest flights then being made. When they completed work on the *Red Rose* a few weeks later, Jessie and Bill still had to load the plane on a trolley to make the sixteen-mile trip from the shed to the racecourse with a police guard ahead of them to clear the road. By the time they reached the racecourse, thousands of local people were clustered around it. So, on 11 March, Bill climbed in, warmed up the engine and without a flicker of hesitation taxied down the runway and lifted into the air. Jessie stood there and watched him circling above them. "Well," she thought. "That's that then. We're off again."

Amid public admiration that Jessie was willing to get back in that airplane, she and Bill finally took off from Singapore once again at seven in the morning of 13 March, ignoring any superstition about the date, although Jessie observed later that Bill had omitted entering the date in their logbook. She suspected that when only thirteen of fourteen invited guests had attended the Aero Club lunch just before they had left Singapore prior to their accident, Bill may well have invited the hotel manager to take the vacant chair and thus change the number.[30] Jessie was reluctant to return to the scene of their accident at Muntok, but they had promised the people there who had done so much for them at the time they would return, and so they followed their previous route south once again. This time the grass had been cut on the landing field in their honor, but that night Jessie was kept awake by the monotonous drumbeats and dismal music of a nearby Chinese funeral ceremony and for her it didn't seem a good omen, especially in that place. They next morning, Jessie didn't dare to breathe as they took off at exactly the same place where disaster had previously intervened until they had passed that fateful 150-foot elevation point. They didn't fly their usual farewell circuit but headed straight out to sea, cheering and slapping the fuselage for luck as they crossed the coastline. At last truly on their way to Australia, they began to island-hop southeast through the Dutch East Indies, now Indonesia, making use of fuel supplies that Shell had set up for Ross and Keith Smith in 1919.

Landing in the rain at Batavia, now Jakarta, they stayed overnight with the British Consul before leaving at four in the morning for Surabaya, flying over a sea that Jessie remembered looked just like curdled oil. As they neared Surabaya, where there was a naval and seaplane base at the mouth of the Madura Strait, along with an aerodrome, a pair of seaplanes flew out to meet them and escort them in, and on landing they were officially welcomed by the commandant and a number of officers with a speech and a presentation to Jessie of a large bunch of red roses tied with a long trail of scarlet ribbon on which were printed wishes for good luck. Leaving at dawn again the next morning, they flew through dense clouds and heavy rain, following the eastern coasts of Bali and then Lombok, flying by the active volcanoes of Gunung Rinjani and Baru Jari, at 12,224 feet the second-highest volcano in Indonesia.[31] If there was a gap in the clouds, the two aviators might have been able to see into the wide, oval caldera with its blue lake as they flew by on their way to Sumbawa, skirting the western side of the even bigger smoking caldera of Mount Tambora before battling up over rugged mountains over 4,000 feet high as the temperature steadily climbed and the weather worsened. When

Tambora had exploded with apocalyptic force in September 1815, an entire mile-high mountain peak had been thrown into the stratosphere with such violence that the fallout changed the global climate for three years.[32] Jessie and Bill struggled to find a way through smoke, looming black thunderheads and lightning, occasionally fearing they were lost among volcanic peaks and valleys before they flew out to sea over Sangar and round the coast to the port city of Bima. There, while being cared for by a hospitable Dutchman, Jessie encountered the largest and hungriest mosquitoes she'd ever seen and was consequently plagued by recurring bouts of malaria for months.[33]

As the sun rose the next morning, Saturday, they took off for Atambua on the island of Timor, where the airfield is about 2,000 feet above sea level, surrounded by mountains towering 5,000 feet higher still. There they descended to find the whole place underwater and covered in long grass through which a narrow strip had been cut to enable them to land safely. As they climbed stiffly from their cockpits, they were thrilled that this would be their last stop before the final hop to Australia. For Jessie, all the hardships and perils of this flight were forgotten in that moment when she felt so near to her homeland. Although it was fiercely hot, Jessie and Bill got stuck into their engine maintenance as soon as the official welcomes were over. From here on they would be flying over the Timor Sea and there'd be no room for error. By late in the afternoon, however, they were forced to shelter under the airplane by the teeming rain. Finally they had to admit defeat, cover the plane and hope the rain would stop by morning. Fortunately, it did and the ground rapidly dried. Jessie and Bill asked the Dutch commander if more grass could be put down on the strip to make it easier to take off, and he promptly ordered up a work gang from the local prison. In no time at all the entire strip was cleared and the grass trodden down to form a firm matting as the runway surface. Jessie suspected that there probably wasn't a native grass hut with its walls intact for some distance around them by now, but it worked and they left the ground with only a few yards of runway to spare.[34]

Heading upwards into the mountains and across the island, the *Red Rose* left the last white beach behind as they flew out over the aquamarine Timor Sea southeast towards Darwin. Jessie and Bill had estimated there were 575 miles of open water to cross, putting them out of sight of land for at least four hours, no small feat in those days in a plane of that size with none of the modern features such as radio, radar, or direction-finding equipment. Once out over the water, they flew into heavy rain that obscured visibility. Even when it cleared at intervals, Jessie could see only empty ocean without even a passing ship in sight. It was eerie, and after a while the vast watery nothingness began to get on her nerves. Sure enough, true to form for the *Red Rose*, exactly halfway across the Timor Sea the engine started to miss and pop. Jessie jerked around and gestured a question to Bill, but he just pen-

A fashionable and very modern-looking Jessie Miller accompanied by a rather uncomfortably suited Bill Lancaster alongside the *Red Rose* after their flight from England to Australia, having apparently landed on a beach somewhere (courtesy Dominic Winter Auctioneers).

ciled a note and passed it over: "I don't know. I am afraid she won't stay the course. But anyway, we've done our best."[35]

Inevitably, despite all they could do, the sputtering *Red Rose* gradually lost altitude as the engine kept sputtering until they were only a few hundred feet above the water. The engine would pick up for a while and they'd gain height; then it would sputter and miss and they would lose height again. This went on for hours. They would discover later that Lancaster had nearly killed them again: he'd forgotten to turn off the tap on the oil reserve and the plugs just kept oiling up. They'd clear intermittently and the engine would run well and they'd gain elevation, but then the plugs would oil up and the engine would sputter and the plane would descend. The *Red Rose* would sink so low that, to Jessie, she seemed close enough to touch the sharks clearly visible beneath her. Jessie sighed and began to mentally prepare herself to ditch, wondering if they could tie the air cushions and life jackets together and float until land was sighted. Otherwise, their only hope was to reach land before they ran out of air. Flying blind as it began to rain, Jessie didn't realize that their miracle had actually happened until they were practically on top of the coast. After some seven and a half hours of worry, Jessie turned around and gave Bill a joyous thumbs-up, and later she always reminded people that, whatever Bill's faults might have been, he had flown hundreds of miles in bad weather across open sea in the cockpit of a small biplane, with only a compass and chart as navigation aids, to arrive over the coast of Australia within one mile of Darwin.[36]

However, as they circled the local airfield at Fannie Bay, just outside of town, they could see it was not only underwater but also completely empty. There was no one to meet them at all, not even one person down there who could indicate if it was safe to land. Not that they had any choice: they were about to run out of fuel. They could really only tell it was a landing ground because fortunately the superintendent of the Fannie Bay Gaol had laid out strips of white calico to mark it and tied red bunting to the telephone poles and wires to render them visible to pilots.[37] So, Bill was able to bring them down carefully in a shower of water at two-thirty in the afternoon of Monday, 19 March 1928. Climbing wearily out of the *Red Rose* onto a completely waterlogged airfield in pouring rain, Jessie and Bill looked around at an empty tropical coastal field bordered by a prison, then looked back at each other, both of them equally wet and bedraggled. Bursting into laughter, they shrugged and shook hands. They had just completed the longest flight that had ever been made by two people in a small aircraft of that size, during which Jessie had become the first woman to travel by air between England and Australia, the first woman to cross the equator in the air, and the first woman to ever travel so far in an airplane.

"My God," Bill grinned. "What an awful landing ground."[38]

Four

Australia Fair

It was in fact because of the appalling weather that there was no official welcome party at Darwin. Jessie and Bill had wired Darwin from Atambua that they were on the way, but a reply warning them not to attempt a landing because the airfield was underwater did not arrive until after they left. Their arrival had previously been expected on Sunday afternoon and hundreds of people had endured bad weather to wait at the open field near the Fannie Bay Gaol, only giving up late in the afternoon. Fortunately, by the time Jessie and Bill came in for their unexpected landing on Monday afternoon, the water had gone down a little, but it was still over their ankles when they climbed out of the plane and jumped onto the ground. Desperate for a cigarette, they went through all their pockets but couldn't find a dry match. Just then someone, possibly Mr. Dempsey, superintendent of the Fannie Bay Gaol adjacent to the airfield, appeared on the other side of the airstrip, but they didn't walk over. Puzzled, Jessie and Bill called to them and asked if they had any matches, but instead of approaching they stayed where they were and shouted that they would have to throw the matches to Jessie and Bill because no one was allowed near them until the health inspector had arrived and checked them for contagious tropical diseases.[1] The matches were tossed and caught, and the aviators lit up gratefully. Gradually, more people drifted onto the edges of the airfield, but still no one came near them until the inspecting doctor arrived and checked their paperwork, verifying that they had indeed been inoculated for each and every fever. It had not been quite the welcome to Australia they had imagined.

However, the doctor was quickly followed by a fleet of cars that included Mayor Watts, Customs officials and other government medical officers. Of course, some of the first news they heard was that Bert Hinkler had reached Darwin almost a month before them and had been touring the larger cities. Hinkler had actually cut their time in half: his actual flight time was just under sixteen days while theirs had been thirty-two. While their flight had cost some £2,000 of their own money, Hinkler had been predominantly spon-

Jessie Miller and Bill Lancaster with the *Red Rose* after landing in Darwin, January 1928. Jessie is wearing the shorts for which she quickly became famous (or infamous, depending on the observer's conservatism) and which started an Australian female fashion trend (photographed by V. Fletcher. Used with permission Northern Territory Library).

sored. On the other hand, he had more experience at long-distance flying, was better equipped, and although he'd also flown an Avro Avian, it had been modified for long-distance flight and better maintained. He'd had none of the mechanical problems that had plagued Jessie and Bill and no accidents. Eventually Jessie and Bill and the *Red Rose* passed all the inspections, and the two aviators were loaded into a car and driven in to Darwin for an official afternoon tea reception. There the Mayor of Darwin handed Jessie a congratulatory telegram from Mr. Gould, manager of Shell Australia, which had supplied fuel for the flight. They liked to make the point that the first woman to fly to Australia had used fuel from the unused supply depots Shell had established for Ross and Keith Smith, the first airmen to fly to Australia. "Her success in being the first woman to fly to Australia," the telegram stated, "is a fitting culmination to splendid achievements [of] her countrymen: the Smith brothers, Parer and McIntosh, and Hinkler. Australians will be genuinely proud of this exploit and warmly welcome Mrs. Miller to Australia."[2] And there lay the rub, as Hamlet would have said.[3] From now on, because Bill was not the first man to fly solo to Australia, not even in a small plane, it was Jessie Miller who instead became the noticeable focus of public attention. As well as fulfilling her own ambition, of course, one of the reasons Jessie had been included in the flight had been because the presence of a woman would provide potential for additional publicity about Bill's flight, but now that his achievement and his role had been superseded by Hinkler, this flight had become all about Jessie. Now she was the one who had broken flight records. As he was in the *Red Rose*, Bill was now relegated to the backseat while Jessie became the heroine of the moment.

Early in the afternoon, Jessie and Bill returned to the *Red Rose* and attempted to carry out their usual maintenance ritual and to ensure the plane was safely secured for the night. However, as it was the wet season after all, they were interrupted by torrential rain throughout the afternoon and ended up being late to their own public dinner reception. Having changed from her customary open-neck white shirt and shorts into a dress, the tiny bobbed-haired Jessie was mobbed by an adoring press while she puffed on a cigarette. Despite all that time in an open cockpit, she had "the complexion of a rose," the *Brisbane Courier* journalist gushed, making a heavy-handed connection to the plane.[4] They weren't the only one to do it: Jessie consistently received bunches of red roses throughout their flight. The Mayor announced that Darwin was proud to welcome the first woman to fly from England to Australia, and he observed that their flight had pioneered a route for an airmail service. He read out a telegram of congratulations on the completion of their flight to Australia "on behalf of the Government of the Commonwealth and the people of Australia" from the Prime Minister, Stanley Bruce, who added, "We are grateful that despite the difficulties and delays which beset you on the journey, your task has at length been safely accomplished."[5] However, he informed reporters, it would be unlikely those congratulations would extend to an official welcome to Canberra. Unlike the political adoration extended to Hinkler, there'd be no invitation from the Speaker to Jessie and Bill to be seated in the House of Representatives. Nevertheless, in reply to the Mayor's speech Jessie graciously thanked everyone for their warm welcome, humbly declaring that she'd never made a speech in her life and had no idea what she should say. She'd soon be more accustomed to speeches than she'd thought possible.

The next morning, they again returned to the plane and with the help of former air forces flight sergeant Gordon Cowper continued their thorough overhaul of the engine. Cowper would later remark on the excellent condition of the engine, considering the aviators had been largely maintaining it themselves. "Mrs. Miller," observed a local journalist, "is quite a good mechanic, wielding a spanner joyfully."⁶ A sign writer was located to repaint the sun-faded and sand-blasted letters spelling out *Red Rose* along the sides while reporters arrived to ask Jessie and Bill silly questions. Yes, Jessie was forced to reply with tongue firmly in cheek, she was longing to see her husband. The two had to pose for photo after photo, eventually causing Jessie to observe tartly that "they must have all the cameras in the world in Darwin right now." Informed of the town's population figure, she commented that she must have met that many people over the last twenty-four hours.⁷ Meanwhile, the Mayor was reassuring the press that, if necessary, Aboriginal prisoners could be used to spread gravel on the soggy airstrip to ensure it was safe for takeoff. In the end, they could only make a narrow strip suitable for takeoff given the soggy condition of the field.

So, with the added difficulty of a crosswind that reduced their takeoff speed to only 45 miles per hour, a rejuvenated *Red Rose* lifted above Darwin at eight in the morning on Thursday, 22 March. After scaring some inhabitants as Bill dropped low over the roof of Mayor Watts' house, he pointed the airplane's nose southeast towards Newcastle Waters, Longreach and Charleville, from where Jessie and Bill would follow the regular Qantas airmail route south.⁸

Meanwhile, southwest of them and out in the middle of nowhere, Jessie and Bill discovered that maps supplied to them by the Civil Aviation Department in Darwin were inaccurate and useless, as Bert Hinkler had discovered before them when he'd actually become lost trying to follow them having successfully navigated himself halfway around the world. Some twenty-five miles out from Newcastle Waters, Jessie was craning her neck over the side of the plane, looking for a windmill marked on the map that was nowhere in sight, when she saw two stockmen below them on horseback. Deciding they might as well ask the men for directions, the fliers brought the plane down and landed near them. As if planes came out of the blue outback sky every day, the two drovers were so phlegmatically unconcerned that they didn't even bother to dismount but just waited there in the heat as Jessie climbed down and strolled over. When the older of the men asked where they'd come from, Jessie replied: "London." He just nodded, tilted the brim of his hat and commented that he'd also traveled lately, some 500 miles on horseback in five months. At the moment they were riding to the nearest doctor, still days away, as the younger of the two men was ill. Bill immediately offered to fly the sick stockman to his destination, but he just shook his head and said he would prefer to ride in on his horse. The other drover looked at their map and pointed out some landmarks that would guide them on the way to Newcastle Waters and then Jessie climbed back into her cockpit, having been brought back to earth by meeting a man who preferred to keep his feet on the ground, or at least his seat in a saddle.⁹

From Newcastle Waters, they flew southeast the next morning to Brunette Downs on the eastern side of Lake Sylvester, where they were forced down at noon by the worst heat Jessie thought she'd ever experienced, and she came from Southern Cross! They had tried ascending to 5,000 feet, but it only seemed to get hotter. A few hours later, accompanied by the Qantas mail plane, they took off for Camooweal on the Barkly Tablelands, where they stopped for the night to refuel and rest. About 188 km northwest of the mining town of Mount Isa and only 12 km over the Northern Territory border, the tiny town of Camooweal was their first stop in Queensland. From there they flew on Saturday to the mining town of Mount Isa, where they stopped briefly before flying on to Cloncurry, where Jessie and Bill were met by a large cheering crowd as they arrived, followed by a breakfast reception by the local council at the Post Office Hotel.

They took off again for Longreach but were again forced down by heat, thirst and a very stiff headwind at Winton. Mr. Wellington, a shearing contractor at Julia Creek, happened to be shearing near Winton when he heard the low-flying plane and saw it overhead. Jumping into the saddle, he raced the plane to the airstrip to meet the fliers while they flew over him, waving. Another enthusiastic crowd was waiting for them on the ground on 24 March as Jessie and Bill arrived over Longreach at six in the evening, accompanied by a Qantas plane that had met them some twenty miles out. As soon as they landed, Jessie jumped out and hung on to the wing to help the plane turn sharply and taxi into the hangar. They were officially welcomed by Roy Petersen, chairman of the Shire Council, who called for three cheers. Taking advantage of a cozy, well-lit hangar, Jessie and Bill decided to wait until later in the evening to work on the plane. Meanwhile, they were guests at an official dinner hosted by Shire Chairman Roy Petersen and Mrs. Petersen at their home, along with others who included the managing director of Qantas, Hudson Fysh, his wife and the secretary of Qantas, E.D. Miller, and his wife. Then it was back to the hangar and working on the Avro until midnight, with some help from Shell staff and the Qantas chief engineer. The engine was completely overhauled, including fitting a new set of spark plugs, but it was still not performing entirely to Bill's satisfaction the following morning. He had Jessie lie across the fuselage behind his cockpit to keep the tail on the ground while he wound the engine revolutions up as high as he could. By then a crowd was gathering, even though it was only six in the morning. While Jessie commented that she thought it awfully good of them, Bill wondered whether they were secretly hoping to see them crash. As they took off, Hudson Fysh accompanied them into the air in his own private plane, flying out with them for about twenty miles, circling while he watched them fade into the clear blue distance.[10]

By now, it was even clearer whom the media favored. Bill might be a competent pilot, but it was Jessie who was the first Australian woman of the air. Already she was under contract to an Australian newspaper for her story, and she'd received an offer of £250 a week for a lecture tour in Australian theaters. Lady Enid de Chair, wife of the New South Wales state governor, told reporters that she had "the greatest admiration for her courage and enterprise" and wished she had been able to make the same "sporting effort" as Jessie. "It was another proof that women could more than hold their own when it came to courage and endurance," she concluded. A leading socialite, Mrs. Albert Littlejohn, went further and declared that "the women of Australia should give Mrs. Miller a tremendous welcome" because she had "put women on the map."[11] The state president of the Country Women's Association, Mrs. H. Griffiths, congratulated Jessie on her courage and predicted that "women in future may follow where she has not feared to lead."[12] They now had a deadline for their flight, too; the Aero Club of N.S.W. had invited them to arrive in Sydney on 31 March to enable them to participate in their aerial pageant of thirty airplanes on 2 April. Inevitable comparisons were already being made and debated between the flight of the *Red Rose* and that of Hinkler, but neither Bert nor Bill would ever say a critical word about each other. As a fellow Avro pilot, Hinkler had been among the well-wishers to see Jessie and Bill off from England and Bill had of course been there to meet and help Hinkler in Singapore. They were all fellow aviators adventuring through the same skies and sharing each other's problems; the men considered themselves part of a brotherhood of the air and considered it against their code to speak ill of each other. "Hinkler's flight," Bill declared in Brisbane, "is probably the greatest British flight that has ever taken place." Jessie, on the other hand, in her usual outspoken and candid manner, was always less inclined to be so generous about their competitor. As far as she was concerned, "That Hinkler butted in before us which, of course, he could not have done but for the crash."[13] Jessie considered that Hinkler had taken advantage of their situation to steal the prize from them that they had struggled so hard to achieve. Bill would never have this much fortune again, and so when Jessie reflected on those events later it must have seemed to her that both their lives might have flown a very different course if Bill had achieved his goal of being the first man to bring a small plane from England to Australia.

Late in the afternoon of Monday, 26 March, Jessie and Bill arrived at Eagle Farm in Brisbane from Toowoomba via Charleville and Roma, accompanied by three Qantas planes Judging by the hundreds of letters waiting for her in Charleville, Jessie's popularity was increasing by the day. Carried away by their enthusiasm, the Charleville Town Council sent a telegram to Kiki Lancaster back in England congratulating her on the "wonderful feat achieved by your husband" and that this "most worthy performance in the London–Australia route to date" was even more worthy because one of the aviators was a woman.[14] One can imagine that must have really made Kiki's day. Almost the entire population of Toowoomba, several thousand people, had turned out to welcome them as they landed there just before noon, showing their appreciation of the duo's achievements by sounding their car horns. The *Brisbane Courier* journalist observed that Jessie, dressed in her trademark white shirt and shorts, looked wonderfully fit and that the flight of Bill and "his brave little companion had fired the imagination of the people," who marveled that this woman who could refuel planes and spin propellers was "quite as competent in her duties as the pilot."[15] Showing off her competency just a little for the huge crowd, Jessie stood on the wing as they taxied into position for takeoff before climbing down into her cockpit. Jessie must have had mixed feelings about being in Toowoomba again. Some three years previously, she'd stayed there for nearly two weeks while traveling to Roma with her sick brother, who had eventually passed away. Even revisiting the town briefly must have wakened memories for her.

In Brisbane, the wildly cheering crowd broke through the barriers and ran across the field to meet Jessie and Bill as the *Red Rose* touched down, only to be restrained just short of the spinning propeller by mounted police. There Jessie and Bill were officially greeted by the Mayor, Alderman W.A. Jolly, the Australian manager of the Shell Company, Mr. Smith, and Ross and Alan Philp from C.C. Wakefield. Everyone, another intrepid reporter

Jessie, Bill and the *Red Rose* landing at Toowoomba, Queensland, 26 March 1928, on their way south to Brisbane and ultimately Hobart, Tasmania (courtesy Howard Rub and the Darling Downs Aero Club).

Looking the image of the tropical aviator, Jessie in a fashionable sun helmet, having landed at Toowoomba, has just been presented with a giant floral bouquet by the Lady Mayoress behind her (courtesy Howard Rub and the Darling Downs Aero Club).

Jessie and Bill refueling at Toowoomba just before taking off for Brisbane (courtesy Howard Rub and the Darling Downs Aero Club).

observed, was "eager to see the plucky little woman who has earned the distinction of being the first of her sex to fly to Australia." Ever inclined to focus on her looks as much as her abilities, the *Brisbane Courier* journalist pointed out that Jessie "looked more like a charming boy than a woman who has been married for seven years" with her cropped dark hair, "a brilliant complexion and laughing dark eyes," and reported that all the women had wanted to know the skin cream she used. Jessie just retorted that there'd been no room in the plane for more than a bar of soap. The "plucky little woman" must have been just as happy with the comment that everyone agreed "her nickname 'Chubby' described her in a word."[16] The journalist from the *Telegraph* became even more carried away about her small build, her large brown eyes, her slight tan, her outfit of aviator's cap and goggles with white linen shirt and shorts, matched with brown golf stockings and brown kid shoes on "her extremely small feet," marveling again that Jessie had come all this way with only a small bag as luggage. At the dinner reception at Lennon's Hotel that evening, Jessie was transformed into an elegant lady in ivory silk who chatted happily to the women of the press with cigarette in hand, but perhaps much of that was just for show. "Sometimes I wish I had never done it," she reflected to them. "I really feel scared stiff and want to creep away to bed.... I thought we would arrive just quietly, you know."[17]

On Wednesday morning, Jessie and Bill were given the full official civic reception at the Town Hall hosted by Mayor Jolly, who declared, "These two heroes are making history—history of which every Australian will be proud." The room was decorated with Queensland ferns and palms as well as red roses, of course, which Jessie was also carrying as a bouquet, with the Union Jack and the Australian flag displayed behind the podium. The Mayor congratulated them on their achievement, commenting that one could not help but be impressed with their perseverance and by the spirit of adventure that had inspired them, a spirit that characterized the British race and had resulted in the British flag being planted in so much of the world. He then singled out Jessie, as the first Australian woman to fly from England to Australia, for special congratulation. Jessie modestly replied that she had never been so nervous in her life and felt rather silly doing this. but she was still very grateful for everyone's kindness and was very proud to be an Australian.[18]

Meanwhile, preparations were under way in Sydney for a big welcome there as part of the New South Wales Aero Club Pageant. The *Red Rose* would be met on approach to Sydney by an escort of thirty planes at about five on the Saturday afternoon by way of providing a grand finale to the day's events. The Pageant was already being advertised as the greatest gathering of airplanes then seen in Australia. During the afternoon the Prime Minister, Stanley Bruce, would formally accept the Sir Charles Wakefield Gift airplane in honor of Bert Hinkler, an Avro Avian piloted by Captain Edgar Percival.

There would be an Aerial Derby with twenty-three competitors, including Major H. De Havilland, who would fly from Melbourne in his biplane *Prudence II*. Yet, in the face of all this public acclaim, when asked in Parliament whether an official invitation to Canberra had yet been made to Jessie and Bill, Prime Minister Bruce said he was still thinking about it. It didn't go unnoticed, either, that the state leader Premier "Big Bill" McCormack had been conspicuous by his absence from any of the official welcome functions, along with any member of the state government, compared to their eagerness to be on show with Bert Hinkler. People were beginning to wonder if this reluctance to give recognition on behalf of politicians was because Jessie was a woman.[19]

Meanwhile, Jessie and Bill took advantage of every public relations opportunity in Brisbane. On Tuesday night they were heard on radio in Australia for the first time, advocating the ease and future popularity of flying on station 4QG, and they spoke at a series of dinners and receptions held at various clubs and organizations. Early on Friday morning, Jessie notched up another first: she became the first woman to be given a reception by Brisbane's Stock Exchange Club in their premises. After several long speeches by members, Jessie thanked them warmly and said she fully appreciated the honor she'd been given. Then she and Bill drove back out to Eagle Farm, climbed back in the *Red Rose* and took off for Grafton, where they stopped briefly for the usual official welcome before continuing on to Newcastle, which they reached with only ten minutes of fuel left in the tank. As the *Red Rose* touched gingerly down on the waterlogged racecourse, the plane was suddenly hit by a crosswind gust that spun it around at right angles to the fence. Jessie jumped out of the cockpit onto the ground and threw her slight weight against the back of the wing, enabling Bill to spin the plane around and miss the fence by inches. Covered in mud and drenched in rain, she was still making sure her face was all still there when the Mayor, Mr. Gibson, and the official party ran up to them. Understandably, as soon as the official welcome was over Jessie excused herself and retired to her hotel room for a much-needed warm bath and a good night's sleep.

After another civic reception the next morning, she and Bill left for Mascot late in the afternoon. There, to their amazement, one hundred thousand people were waiting for them in wet and windy weather at Mascot Aerodrome in Sydney on that wet Saturday, 31 March, and it was estimated that another 50,000 were watching from outside the fence. According to plan, they had been met some miles out and escorted in by planes already in the air to form the final dramatic event of that day's Air Pageant. Representatives from four states as well as from the RAAF Richmond and Point Cook bases, the Civil Aviation Department and various civilians had taken part in the Air Derby during which twenty-three planes had flown a ten-mile course. There had been aerobatic demonstrations, a dogfight during which some of the crowd thought the loser had really crashed, and an attack on a tank. For the first time in Australia, there had also been an Oaks race of the air between Miss M. Reardon, the winner, Mrs. A.M. Upfold and Miss E.M. Follett. Then the *Red Rose* was sighted on approach and everything came to a standstill as thousands of people craned their necks to see Jessie and Bill land. As a malaria-stricken Jessie climbed wearily out of the plane, her husband, Keith Miller, and Captain Chateau of the Aero Club ran to the plane to greet her. Realizing the plane had come to a halt in mud several inches deep, they formed a chair with their hands and carried her to the open car, while Bill arrived borne on the shoulders of some other people. They then drove around the field in a victory lap while Jessie smiled and waved to the wildly cheering crowd. On the official stand to greet her were Jessie's mother and Bill's father, who proudly watched on as she and Bill were officially welcomed by the Governor and Lady De Chair; William Glasgow, Minister for Defense; Mr. Latham, Federal Attorney-General; Mr. Boyce, the State Attorney-General representing the Premier; Captain Geoffrey Hughes, president of the Aero Club; the Mayor of Mascot, Michael L'Estrange; Sir Keith Smith; and Mr. Cyril Westcott representing Sir Charles Wakefield. Jessie was once again handed a large bouquet of roses and then the two fliers were presented with official addresses of welcome by the Mayor, who congratulated Jessie on her "indomitable courage." Bill thanked everyone, but the crowd called for Jessie, who replied modestly, "I cannot make speeches, and I am too overcome by this welcome you have given us. I can't tell you how glad I am to be back, but I am so tired. I know you will forgive me. The welcome is beyond anything we could have imagined. We thank you from the bottom of our hearts."[20]

After another welcome at the Mascot Town Hall, the two weary aviators finally retired to separate rooms at the Hotel Australia. One can only wonder whether Jessie took this opportunity to explain anything at all about her association with Bill to the patient and long-suffering Keith or, given that it was still a well-kept secret, whether she felt the need to reveal anything at all. Meanwhile Bill's father was taking the opportunity to point out to reporters that the flight had been largely financed by him and his wife, Sister Red Rose from the Mission of Flowers, and that it had the full support of Bill's wife, "a personal friend of Mrs. Miller," as well as the support of Keith Miller and Mrs. Beveridge.[21] What he didn't tell the press was that, having seen which way the publicity wind was blowing, he'd been quietly urging Bill to ditch Jessie and do the rest of the Australian tour solo. Bill had valiantly replied that Jessie had gone through the same hardships and effort as he and he wouldn't dream of abandoning her.[22] Jessie's mother, on the other hand, said that although she had been anxious about her daughter's flight, "I was always convinced that my daughter would get through." For her, it was family history repeating itself.[23]

Four. Australia Fair 77

Jessie, with Bill standing to her right, inhaling yet another rose bouquet on landing in Sydney at the end of March 1928 (author's collection).

On Monday, Jessie and Bill were guests at the Australian Aero Club's dinner, where they were each presented with a £100 check and a handsome commemorative album. However, Jessie got only short and chauvinistic notice about the Sydney event from her home-state newspaper, the *West Australian*. Reminding readers that her pet name was Chubby, the paper reassured them that this pioneer Australian aviator was also a "good, all-round girl who can turn her hand to anything, an accomplished pianist, a good cook, and able to make her own clothes." However, they added a little more astutely, she "was always plucky and if she made up her mind to do a thing, you knew she would go through with it." Though brief, those words speak volumes about the attitudes of the time towards women.[24]

Only when Federal politicians finally realized just how popular Jessie and Bill were in the eyes of the Australian public, that is to say voters, did they eventually begrudge them an official invitation, although neither of them was still considered important enough to meet the Prime Minister face-to-face. Leaving Sydney on 3 April, the fliers landed at Canberra's Review Ground to be welcomed on the Prime Minister's behalf by the Minister for Home and

Territories, Sir Neville Howse. After settling in at the Hotel Canberra, Jessie and Bill were given a luncheon reception at Parliament House. Once again, the press was more interested in what Jessie wore than what she'd achieved, reassuring their readers that she was still a woman (and thus no threat to men). A journalist from the *Canberra Times*, who met Jessie that evening standing alongside the *Red Rose*, observed that she was now wearing a "smartly-cut frock of navy crepe marocain with a neat little felt hat," which was in sharp contrast to "the boyish figure which had leaped out of the plane" that morning. She had "soft brown eyes" and "her features have been richly tanned by the sun and her voice is soft with an English accent." Asked what she thought of the Federal City, the "gallant little lady" got in a return volley at the politicians, who had only moved there in May of the previous year: "I think it is very beautiful indeed, particularly the flowers, but it hardly looks like a city."[25]

Meanwhile, it wasn't all flowers and happiness down in Melbourne, the next major stop for the Lancaster-Miller flying circus. Despite the Aero Club

The *Red Rose* in September 1930, after her sale and later reregistration, probably photographed in Sydney (courtesy Ed Coates and the Frank Walters Collection).

having made elaborate plans for their reception at Essendon Aerodrome on Wednesday, including afternoon tea and a dinner dance, Jessie and Bill had postponed their flight because they had been made an offer by Mr. Campbell, the manager of Melbourne Carnivals Ltd., to land at his Motordrome on Saturday, 14 April. Although they must have known that Hinkler had declined a similar offer because he considered the Motordrome unsafe for an aircraft landing, Jessie and Bill postponed their flight while the matter was negotiated. Jessie had been ordered to rest by her doctor, and so Bill and his father took the unusual step of boarding a train for a personal inspection tour at Melbourne, where they were met by Captain Jones of the Civil Aviation Department, Mr. Campbell, and some executives from Shell. Bill took one look at the venue and could see Hinkler's point. To attempt to land within the track, he'd have to descend too low over the heads of the crowd for his liking; if the engine misfired, it could end in a disaster that would make entirely the wrong kind of headlines. Consequently, Bill refused to do it, backpedaling his way out of the situation by claiming he knew nothing about any official invitation or expectation they would arrive on Wednesday. Instead, Bill directed newspaper reporters to Keith Miller, who he said was their business manager.

Arriving back in Canberra on Friday, Bill met up with his brother J.K. Lancaster, who had also been a World War I pilot and was now manager of the General Electric Company, at his Red Hill home and the next day joined Jessie for their takeoff from Canberra at ten-thirty in the morning. They landed at Holbrook around the middle of the day, where John Ross of Kanimbla Station invited them for lunch. Typically for Bill, and to Jessie's usual irritation, he hadn't bothered to bring a map. "We were 6000 feet up, and how was I to know where North Essendon was?" she snapped on landing. "All I knew was that it was a suburb somewhere."[26] So, they had to land again in someone's evidently very large backyard at Ivanhoe to ask that startled resident for directions to Essendon, where some four thousand people had gathered during the afternoon to welcome them. More than half of them were women. At about three in the afternoon, seven aircraft led by Captain G.S. Matthews in a DH4 set out to meet the *Red Rose* in the air but had to return thirty minutes later empty-handed. Just when the crowd was starting to grow restless and anxious, the little biplane suddenly swept over their heads from the east only a few hundred feet above them. As Jessie leaned over the side of the cockpit and waved, the relieved crowd cheered and waved back. Circling around, Bill descended quickly and touched down, but as the *Red Rose* taxied across the airfield the crowd broke through the barriers and ran after it, surrounding the plane before it stopped, narrowly missing the still-spinning propeller and tragedy. Jessie, dressed this time as a stereotypical aviatrix in jodhpurs, leather jacket, cap and goggles, leaped out of the cockpit and embraced her mother and friends, declaring, "We lost our way and he blames me."[27]

After the plane had been pushed into a hangar, Jessie and Bill were driven through the cheering crowd to the Aero Club building for an official welcome. Still looking very tired, Jessie told the assembled members and guests that she was now so deaf from the engine that she couldn't hear herself speaking. She was officially welcomed by Mrs. A. Glencross, president of the National Council of Women of Victoria and the Australian Federation of Housewives Association. Expressing pride that Jessie had achieved something no woman had before, Mrs. Glencross declared, "I have always maintained that most women can do what most men can do when occasion arises. I am proud that Australia has produced a woman of such courage and capability as Mrs. Miller and we are sure that hers will be a forerunner of other great achievements by Australian women." And so they were. Perhaps another achievement for Jessie was the message from Kiki Lancaster waiting for her: "Three cheers for your success; hearty congratulations. Love, Kiki."[28]

Colonel H.C. Brinsmead, Director of the Civil Aviation Department, was there to be part of the official welcome and had the last word about the landing site dispute; after observing that Lancaster had flown all the way from England in a "tiny little machine, and the fact that he'd been accompanied by a tiny little woman had stirred popular imagination," he calculated that the *Red Rose* had landed at a speed of 90 miles an hour, needing some 280 yards to stop. That, he declared, would have resulted in a major tragedy at the Motordrome, where a plane landing in the center of the track would have had only 100 yards in which to stop before hitting the banked concrete track and sailing into the crowd.[29]

As the following day was a Sunday, Jessie and Bill took time out to relax and went with Keith Miller to the Williamstown horse races, where Jessie backed three winners. The next day they met with executives of Union Theatres and confirmed a contract for an illustrated lecture tour of capital cities, which would begin at the end of April. On Wednesday, however, it was back to a long day of official duties beginning with a civic reception in the morning at the Town Hall hosted by the Lord Mayor, Sir Stephen Morrell, that included representatives from women's organizations, the Royal Australian Air Force, and the Civil Aviation area. The Mayor outlined the main highlights of their trip in his speech, emphasizing their perseverance and courage, and concluded by stating that Australians were proud an Australian woman had been the first to fly the distance. Colonel Brinsmead also emphasized their courage by reminding the audience of recent flying tragedies involving Princess Lowenstein and Elsie Mackay.[30] Few young people had the chance to achieve such immortality as Lancaster and Miller, he said. In reply, Bill informed the audience that the first congratulatory telegram they had opened at Darwin was from the Lord Mayor of Melbourne. Jessie had declared that Melbourne was really where she'd been heading for all the time and apologized for having the misfortune not to be born in Victoria, as she felt that Melbourne was her home.

From the Town Hall, they then moved two blocks northwest to Menzies Hotel, where the Shell Company of Australia hosted an official lunch reception attended by a similar cast that included the Premier, Mr. Hogan, the Lord Mayor once again, the president of the RSL, Mr. K.K. Turnbull, and the acting General Manager for Shell, Mr. W. Gould. With more than a little insight, although few appeared to take him seriously, Mr. Gould suggested in his opening speech that Mr. Hogan should consider the airplane might soon be in competition with the railways and that instead of building steam engines at the Newport workshops, the Railways Commissioners should be building aircraft engines there. Prophetically, the Newport workshops have since vanished into history; perhaps Mr. Hogan should have listened more carefully that day. Gould concluded by reminding the gathering that Jessie had now established a precedent for women in aviation. Mrs. Glencross reemphasized Jessie's achievement with another transport metaphor: she may not have left footprints on the sands of time, but Jessie had blazed a track through the air, Mrs. Glencross extolled, and so she considered it a great honor to welcome Jessie on behalf of the women of Victoria. An overwhelmed Jessie humbly thanked everybody, unable to resist telling them that when Bill had told his father he was taking her on the trip Mr. Lancaster had replied, "Ridiculous nonsense! No woman could stand it. Besides, she's bound to get ill." An embarrassed Bill protested to the room that of course Jessie had done a great deal of work with and on the plane for which he was very grateful. Then he broke the news that they would be embarking on a lecture tour around the country, beginning soon in Melbourne, and that they would be only too happy to record the tour on film if asked to do so. They had in fact signed a contract with Union Theatres to fly to Australian state capital cities and some of the major towns on an illustrated lecture tour from around the end of April. Then in the afternoon it was off to Anzac House at the top of Collins Street to speak to the RSL because of Bill's initial service with the Australian Flying Corps in World War I.[31]

The public relations engagements kept on stacking up, but after all, their financial future could well depend on events like these. On Friday, Jessie and Bill were at a lunch hosted by the Executive of the National Women's Council of Women of Victoria. This time the Lady Mayoress, Lady Morell, and Mrs. Beveridge were among the guests. The redoubtable Mrs. Glencross again seized the chance to commend Jessie and Bill for lessening the distance between the far-flung corners of the Empire. Jessie replied that she hoped their flight could convince people that the danger of flying was not as great as some feared, in fact no more dangerous than crossing the street, and that she also hoped she and Bill could reduce the distance between far-flung corners of Australia. They had brought a letter with them from Darwin, she pointed out, that had reached its destination in only a few days rather than the current four weeks. Before they left Melbourne, Jessie and Bill presented the first of their touring lecture series, illustrated with lantern slides, at the Majestic Theater on the night of Saturday, 21 April. To set the tone for the evening a tiny model of the *Red Rose*, suspended from the ceiling by wire, circled over the heads of the audience while picked out by a spotlight. Introduced by well-known actress Louise Lovely, the two aviators alternated as speakers, dressed in their flying outfits, while colored versions of their photographs of overseas cities and peoples, of sunset skies and banks of clouds, flashed onto the screen behind them.[32]

Towards the end of the month, Bill's father left Melbourne to do some advance publicity across Bass Strait in preparation for their next flight. Because it would be a continuation of their flight down the Australian east coast, someone had realized that Jessie and Bill could claim that their flight had actually begun in England and ended at Hobart in Tasmania. In this way, despite Hinkler, they would be able to claim a record as the first pilots to fly an aircraft from England to Tasmania. Arriving in Launceston on 26 April, Mr. Lancaster didn't hesitate to point out to journalists that his wife had been a major financier of the venture and that the plane was named in her honor. He gave a quick summary of the highlights of Jessie and Bill's flight to Australia and reassured the press that they had attempted to dissuade Jessie from undertaking such a dangerous flight but that she had been determined to do it. He was also careful to point out that Kiki Lancaster was a personal friend of the courageous Jessie and fully supported her being part of the flight.[33]

Despite Bill's good intentions, the two aviators still found themselves the subject of controversy, although one of a different kind than they might have expected: their schedule had them landing at the Elwick Race Course in Hobart on a Sunday. Now, public entertainment on this scale on the Day of Rest was irritating enough to conservative church leaders in that island city, but when the manager of His Majesty's Theatre, F.B. Mason, applied to the council to actually sell tickets at the gate—well, all hell broke loose, so to speak. As if by a miracle, the various denominations suddenly forgot their sectarian claims to each be God's only children and became as one right-wing Christian family body declaring unanimously that such financial transactions on the Sabbath were not only unholy but also undemocratic. As businesses were at that time legally closed on Sundays, someone allowed to sell tickets to an event would have an unfair financial advantage. Besides, to permit ticket sales on a Sunday would create a dreaded precedent; one never knew how far this thin end of the wedge might be driven. Needless to say, Mason's application was profoundly rejected by the God-fearing councilors seven votes to one.

So, some twenty-five thousand enthusiastically cheering people were able to greet the *Red Rose* for free at Elwick on that calm, fine Sunday of 29 April. Well, all except Mr. Mason perhaps, who was probably quietly weeping as he counted heads and thought of the difference a day could make to one's financial future. Our two fearless aviators were accorded a reception, gushed the *Mercury* newspaper the next day, "such as Royalty alone has previously commanded in this city." Five special trains had been run from the city to the track to carry the people, while others came by tram or by various kinds of motor vehicle that were estimated to cover several acres nearby. After twice circling the ground from the north, a little lower each time with the sun glinting off the wings, Bill and Jessie landed at three-thirty and were then driven around the course in a victory lap, acclaimed by thunderous cheering, to where they were welcomed by the Warden of Glenorchy, Mr. Ben Watkins, representatives of Union Theatres and the RSL, and of course the ever-present Mr. Lancaster. Among the usual plaudits, including a telegram of congratulations from the Premier of Tasmania, Joe Lyons, the Warden reassured everybody that it was only fitting, considering they were still carrying and distributing the "Kindness" pamphlets given them by Bill's mother, that they should be welcomed on a Sunday. Bill, welcomed as another "digger" by the RSL, replied that he and Jessie were overwhelmed by the size of the welcome. After briefly settling into Hadley's Hotel, Jessie and Bill visited the local Repatriation Hospital to talk with patients, and then they were guests of the RSL at dinner, where Bill was made a member of the Hobart branch.[34]

Jessie and Bill had flown to Hobart from Launceston, where they had landed the previous day at Mowbray in front of 3.000 people after Jessie had piloted the *Red Rose* for six and a half hours on a 370-mile semi-circular route across Bass Strait and along the north coast of Tasmania through very heavy weather, becoming the first woman to pilot a plane across it.[35] Wearing life belts over their flying jackets, they had left the Essendon airfield at 10:20 am, waved off by Jessie's mother, a representative of the British Imperial Oil Company, the manager of Union Theatres and Louise Lovely. Carrying enough fuel for eight hours as a precaution, they flew west to Warrnambool, then turned south across the Strait for King Island and then Burnie to fly back east along the coast of Tasmania to Launceston, arriving ninety minutes overdue to be welcomed by some 2,000 patient people. Once again, as the journalist from *The Examiner* observed, rather than for Bill all eyes were for the "woman in the case," who, they were apparently glad to see, "was not the least suggestion of the Amazon, but a little lady in leather coat and tight cap, her cheeks stung scarlet by the wind, her eyes dark brown and humorous." They were also pleased to note later that "her skin was unblemished either by bleak starry altitudes or the discomfort of flying and hard work in tropic deserts," and that she had undertaken the flight "simply because she saw nothing to be afraid of."[36] After leaving Launceston in the afternoon, they had followed the railway line south to Hobart. For the first and last time, because she had wanted to look good for the huge crowd at Hobart, Jessie had actually worn a dress and silk stockings in the cockpit, but when Bill had to climb to 10,000 feet to avoid clouds she nearly froze solid. She subsequently vowed never to fall victim to aerial vanity again. However, probably unknown to Jessie, another item of her clothing was breaking fashion boundaries farther north in Australia: her shorts. Because it had been so hot up there, Jessie had worn them in and out of the plane and been consistently photographed in them and women had happily seized on them as a fashion trend, especially on the east coast beaches such as Coolangatta, although some dispute had evidently taken place over whether they were allowed to wear the shorts indoors for hotel dining.[37]

Jessie at this time was really at the forefront of the developing relationship between flying women and fashion. When Amy Johnson made the first solo flight by a woman from England to Australia in 1930, she left England in a full flying suit but, perhaps inspired by Jessie, had also abbreviated her outfit by the time she reached Singapore to shorts, socks, oil-stained men's brogues and a drill jacket, under which was a purple blouse, topped off with a topee. Like Jessie, her short hairstyle, known as the "Amy Johnson wave" or the "Johnnie shingle," was much commented on and copied. While she, too, could service and repair her own plane, having been the first woman in England to earn a Ground Engineer license, for most of the time she was in Australia it was her clothes, her hair and her complexion that newspapers tended to focus on rather than her achievement. Even back in 1911, Harriet Quimby, the first woman to fly across the English Channel and the first woman to be granted a pilot's license by the Aero Club of America, observed, "If a woman wants to fly, first of all she must, of course, abandon skirts and don a knickerbockers uniform. I speak of this particularly because so many have asked me about my flying costume."[38] Jessie was a knickerbockers girl, too, but it was her choice to wear shorts that made her famous, on the one hand declaring her independence but on the other hand baring her own unique type of sexual allure.

On the Monday morning it was off to the Hobart Town Hall reception hosted by the Mayor, J.J. Wignall, and the Mayoress. Despite rain, several hundred people had gathered outside the Hall and the Mayor took political advantage of the occasion to point out that Bill and Jessie had done what the Commonwealth government had failed to do in laying the foundation for a regular air service from the mainland to Tasmania. Bill reinforced the practicality of such a route in his reply while Jessie once more reassured women that flying was safe, that it was "very snug and comfortable seated in an airplane and that once they tried it they would enjoy it."[39] Then in the afternoon and evening, they presented their first lectures in Australia about their flight from England, repeating them the next day and again on Wednesday. Not to be outdone, Bert Hinkler was right behind them as usual, landing in Tasmania the following week at the same place where virtually the entire welcome and reception ceremony would be repeated for him. He'd actually sent Jessie and Bill a telegram of congratulations on the Bass Strait flight. On their last day in Hobart, the fliers took the *Red Rose* up over the city and harbor, flying low over the Cenotaph at the Domain to drop a laurel memorial wreath attached to which was one of Bill's mother's "Kindness" messages. Landing at Elwick again, Bill picked up Bert Cross and flew over Hobart once more in clear, bright weather to enable Cross to take film footage for his tourism film *Tasmania at Work and Play*. Cross shot over 400 feet of film and was ecstatic about the visibility.[40]

Leaving Hobart behind them on Friday, 4 May, Jessie and Bill circled over the city and suburbs for thirty minutes, distributing more "Kindness" messages on their printed cards into the air as they had done over every town at which they had landed, as requested by Bill's mother. They flew the short distance up the river Derwent to New Norfolk, but while attempting to land on the racecourse there Bill had to turn the plane into a hedge to avoid the crowd running towards them. Landing in rural areas was, he commented later, becoming nerve-wracking. Then they turned north and retraced their flight path for the first time since they left England. They touched down briefly at the coastal towns of Ulverstone and Penguin and then flew over Burnie on their way west to Wynyard, where they landed on the golf course late in the afternoon to be welcomed by the Mr. Jones, the Warden, and a thousand people. They gave a lecture that night and then flew back to Burnie the next day and then on to Scottsdale. They would not, however, be able to create history by being the first airplane to fly over the rugged and remote west coast of the island, as no suitable places to land could be found in time, and that attempt had to be abandoned.

It was then that the odds of flying wood and fabric airplanes into the unknown caught up with Jessie and Bill once more. Landing again at Mowbray in Launceston on Sunday, 6 May, the *Red Rose* was damaged. Spare parts could not be found in time, as Jessie and Bill had to be in Sydney by Friday to fulfill their lecture contract, and so the *Red Rose* had to be lifted ignominiously on board the *Loongana*, on which it sailed back to Melbourne. Meanwhile, Jessie and Bill held three lectures in Launceston over the Monday and Tuesday, enthralling packed houses at the Princess Theatre with their tales of adventure accompanied by projected color slides of the places they visited. They were always at pains to emphasize that they hadn't done the flight as a stunt to gather fame and fortune but as a demonstration that an air service could be established between England and Australia.

An autographed pair of studio portraits of Jessie and Bill to commemorate their landing in Australia, one of the rare occasions Jessie used her nickname Chubbie, which also confirms the unusual spelling (courtesy The Cobbs Auctioneers).

Jessie and Bill returned to Melbourne by ship and then, repairs to the *Red Rose* still not having been completed, caught the express train to Sydney on Saturday. Their repair bill was going to run to over £200 and the plane would still have to be flown up to Sydney after work was finished, so on Saturday, 12 May, they stepped onto the stage at the Crystal Palace Theatre to once more tell their story in all its vividly illustrated detail before a backdrop of blue sky and clouds. So resounding was the applause that they continued on into a second week. At the end of the month, the fliers were invited to a reception given by the Society of Women Painters and there they were welcomed by Mrs. Florence Taylor, wife of flight pioneer George Taylor. It was a rare and remarkable meeting of the two first women of flight in Australia, between the woman who had flown the greatest distance and the first woman to pilot a winged aircraft.[41]

Early in June, the fliers were in Melbourne for a lecture season there, but audiences persisted in being more interested in Jessie rather than Bill. It was Jessie, first woman to Australia in the air, that the crowds really wanted to hear, and she was being paid substantial sums for her lectures, from which she sent Bill's wife, Kiki, £30 a month to help take care of the children while Bill was away. Jessie continued to do this after they went to America until the time came when she informed Bill it was about time he lived up to his own responsibilities and paid for the upkeep of his family with his money, not hers.[42] Then Jessie and Bill flew north, lecturing in a series of towns. They were flying to Canberra on 14 June when a main spar of the left wing cracked and collapsed when they landed at the Cooma racecourse in front of a large crowd. Bill claimed poor materials had been used when the plane was repaired in Melbourne. Rather ignominiously, the pair had to drive into Canberra, where Jessie presented her lecture over the next few days. Further lectures had to be postponed indefinitely, while the *Red Rose* remained at Cooma. Jessie and Bill would never fly it again.[43]

While they were in Canberra, they were invited to join in the official welcome for Charles Kingsford Smith and his crew. Kingsford Smith, co-pilot Charles Ulm and Americans James Warner and Captain Harry Lyon had landed in Brisbane on 9 June after becoming the first to make the transpacific flight from America to Australia in their Fokker trimotor, *Southern Cross*. Bill and Jessie quickly struck up a close friendship with the American duo that would change both their lives.

Five

Burning Bridges

Raised as the son of an American rear admiral, Harry Ware Lyon had been a merchant ship captain until he agreed to become the navigator for Charles Kingsford Smith's record flight across the Pacific, despite having only been up in an airplane a couple of times. Lyon had certainly never navigated one. He in turn had recommended James Warner, a Navy veteran, as radioman. Within only a couple of weeks, the press announced that Jessie and Bill would be leaving for America with the returning Warner and Lyon to do a similar lecture tour of the U.S. as they had been doing in Australia, flying from venue to venue, and that they would possibly be involved in the making of an aviation film. Lyon had perceived that their flight had aroused far more interest in America than it had in Australia, and he could see opportunity beckoning.[1] The four were soon planning a transatlantic flight that would include Lyon and Warner as navigator and radio operator and Bill and Jessie as pilot and co-pilot, after which they would make a flight from Australia to America.

On 23 June, thousands of people crowding the dockside at Circular Quay saw Lancaster, Lyon, Miller, Warner and a pair of kangaroos sail from Sydney on board the Matson Line's SS *Sonoma*, a fast twin-screw 6,279-ton liner that plied the route between San Francisco and Sydney, via Honolulu and Auckland, in nineteen days. While the ship island-hopped across the Pacific, the four aviators had time to talk and various plans for future flights were subsequently born during the voyage along with a baby kangaroo; evidently it had not occurred to anybody to check whether one of their pair of native Australian animals might be female, let alone whether it was pregnant. As the days languidly drifted by, Jessie and Bill became firmly seated on the Lyon-Warner celebration bandwagon. In Suva, Fiji, they were all honored at a civic reception and in Samoa they were welcomed at the wharf by the Governor and his staff, who took them back to his mansion for another reception there. By the time they reached Honolulu, where they transferred to another Matson liner, *Malolo*, the aviators were reported to be planning a non-stop flight from England to New York. After ten military airplanes escorted the *Sonoma* into Honolulu Harbor, where an official committee met the aviators, Lancaster revealed some more details of a plan to fly from Croydon, in England, non-stop to New York in a trimotor Fokker with Lyon and Jessie.[2]

When the *Malolo* docked in San Francisco on 11 July 1928, escorted by boats loaded with leading San Francisco citizens, Jessie and Bill found themselves caught up in the huge civic welcome for Lyon and Warner. Sitting up on the back of their open car, they were driven in a ticker-tape parade through streets lined with cheering crowds to the Town Hall, where they were welcomed by the Mayor and invited to a pre-function party. When he asked Jessie what she'd like to drink, she replied, puzzled, "I thought there was Prohibition in America." Amid roars of laughter from the room, the Mayor swept open sliding doors to reveal one entire wall lined with shelves of bottles and glasses with refrigerated compartments at the bottom for ice. As she circulated among the guests, the women told Jessie that they'd never drunk so much in their lives as they had since Prohibition had been in force.[3] Then they moved on to the Palace Hotel, where there was a banquet reception for 300 people. Lyon and Warner were each presented with checks for $6,000 raised by public subscription and were decorated with the Cross of Malta.

A few days later Jessie, Bill, Warner, Lyon and his wife and mother flew on a Maddux Airline Ford trimotor, which Jessie was invited to pilot for most of the journey, down to Los Angeles for another reception at the Biltmore, where Lyon announced plans that were now for a two-stage long-distance flight from Los Angeles to New York and then London, with Jessie and Bill as pilots and Lyon and Warner as navigator and radio operator, that would take place in the Spring of the following year. If they succeeded in reaching London intact, they would also fly the return leg back to New York. Lyon claimed that contracts had already been signed for the construction of their aircraft, another trimotor of similar design to the *Southern Cross*. However, two days later on 20 July, Warner announced that he'd parted company with Lyon over business and personal differences. Jessie and Bill, who were not as familiar with Lyon's problems as Warner might have been, would soon find out for themselves.[4]

Harry Lyon's father, Rear Admiral Henry Lyon, fell ill in Washington, D.C., in August, and so Jessie flew Lyon and his mother to Washington and then to their home in Paris, Maine. While they were on the way, newspapers broke the story of a rumored bigamy scandal involving Lyon. In the end, though, it was probably some kind of extortion attempt.[5] When they all emerged from the other side of that little storm the plans for a transatlantic flight still seemed intact, and in late September Jessie, Bill and Harry reassured newspaper readers that they still intended to make a round-trip flight in April the following year from New York to London and then from London to Los

Angeles in less than 100 hours. They would fly the route in an 800–950hp all-metal, trimotor monoplane currently under construction with a wingspan of 80 feet. The fuel tanks would be able to carry 2,000 gallons of fuel, giving the plane a range of 4,500 miles at a cruising speed of 100 miles an hour. Bill declared that he remained determined to one day end a similar flight in Australia.[6]

In the meantime, Bill, Jessie and Harry had been preparing to make the first non-stop flight to Bermuda. With the financial backing of well-known explorer and publishing tycoon George Palmer Putnam, they had fitted out an Ireland seaplane, the *Flying Fish*, with a Wright Whirlwind motor and long-range tanks that would take 290 gallons of fuel, enough for eighteen hours of flight even though it wouldn't usually take them more than ten hours to cover the 800 miles. Because of the navigational difficulties of locating the group of islands in the open ocean, especially if the notoriously fickle weather of the region lived up to its reputation, Lyon's expertise would be much needed. At the last minute, however, Putnam decided he wanted to go along on the flight personally and pulled rank, so Jessie lost her co-pilot's seat in the *Flying Fish* for a replacement seat in the backup plane when they all took off from Long Island on 28 October. She must have had a brief chance for a laugh, though, when the trio of intrepid adventurers only made it as far as the coast of New Jersey before they were forced down in shallow water by engine trouble and became stuck fast in mud. But it was only a brief laugh before her own plane developed engine trouble off Hampton Roads, Virginia, and Jessie and her four passengers had to ditch at sea, where they floated for a while before being rescued by the Coast Guard. Eventually the seaplane was retrieved and Bill flew it to the Hampton Roads navy base for repairs, declaring he was still going to fly it to Bermuda with Lyon and Putnam. But it wasn't going to be that year; further engine problems caused them to eventually postpone the trip, blaming water-contaminated fuel. Bill had to content himself for the rest of the year with testing the trimotor engines being built for the transatlantic flight at Curtiss Field, Valley Stream, on Long Island.[7]

As far as Jessie was concerned, Harry Lyon was becoming "a dead loss," as she forthrightly put it.[8] None of his grand plans had yet to actually take tangible shape, and the people from the company in Los Angeles who were building the aircraft the aviators intended to use for the transatlantic crossing now wanted to know their requirements for the fit-out. So Jessie went out to see them and, not wanting to waste any of that time, applied to become qualified as a radio operator while she was there. For this or future flights, it would mean one less person needed in the crew. Another good reason for being on the other side of the country at the beginning of 1929 was Kiki Lancaster, who kept on showing up like the ghost of Christmas Past. For some time now, her letters to Bill had been complaining that she was stuck in a little flat in Fulham, back in England, while Bill was gallivanting around America with Jessie, who was getting all the attention as his partner. Kiki thought it was about her turn for a share of the limelight and so she duly arrived in the U.S. just before Christmas bringing some controversy with her and just a little touch of farce. She was carrying with her the portable compass that Bill had used on the flight from England to Australia in which the needle floated on about four ounces of alcohol, but this was Prohibition and the Customs officials declared that nobody could bring four ounces of alcohol into a country of such pure and total abstinence. The navigation instrument would have to be drained of its offensive liquid, they stated. Bill won his appeal that the compass would then be rendered useless and was finally able to collect it entire.[9] It's quite possible that she had guessed by now that Jessie and Bill's relationship wasn't altogether platonic, and Bill's father may well have suggested to her when he came back from Australia, having been able to observe the couple for a while, that it was high time she spent some time with Bill. Whatever the reason for the visit, that Christmas could not have been happy for either of them: Kiki would have been missing the children and Bill would have been missing Jessie.

During January 1929, Bill and Kiki were joined by Irish-born aviatrix Lady Mary Heath, a long-standing friend of Bill, who had flown to America at the end of the previous year on a lecture tour and to attend an international aeronautical conference in Washington, D.C., as a delegate from the Air League of the British Empire and as the chairperson of the women's section of the League. Heath was a strong supporter of women in aviation. The participation of women in aviation activities, she had said, was "needed for the further development of aviation in this country." Observing that not all women interested in aviation had to be pilots, she pointed to Catherine Wright, who had supported her brothers financially to work on their aircraft. Heath advocated that women who did want to be pilots needed to be physically fit, should be trained with parachutes, should be members of clubs, and should compete in races. In other words, they should immerse themselves in the aviation experience. "In a small plane I feel as though I were a part of it," she declared.[10] Having qualified for a private pilot's license in 1925, Heath had then, in the face of strong male opposition, become the first woman to qualify for a British commercial pilot's license. She followed that by becoming the first woman to parachute from an airplane and, later, the first woman to hold an aircraft mechanic's qualification in the U.S. In 1928 she was the first person to pilot solo an open-cockpit small plane, another Avro Avian like the *Red Rose*, the 13,000 miles from Cape Town to London, nearly dying from heatstroke and gunfire along the way, and then in October she had broken the world altitude record in the same plane, ascending to 24,700 feet. It's an interesting comment on the tightness of aviation networks of the time that Heath later sold that

Avian to Amelia Earhart, who then flew it on a round-trip across the U.S., the first woman to do so. Soon known in the U.S. as "Britain's Lady Lindy," Heath would have a close call with death later in 1929, while flying in the same Cleveland circuit races as Jessie, when her plane crashed through the roof of a factory building, fracturing her skull, within which she would wear a metal plate ever afterwards.

Perhaps to make a point that she, too, could handle flying in a small plane, Kiki flew with Lady Heath down to Miami for the opening of the new airport there. Bill accompanied them in Heath's old Avian, now owned by Earhart, along with the pretty dark-haired seventeen-year-old American aviatrix, Elinor Smith, who later that month would establish a world solo endurance flight record for women by remaining aloft in an airplane over Mitchel Field, Long Island, for 13 hours and 16 minutes. They then flew back to New York to attend an air show where they'd be demonstrating the virtues of the Cirrus engine, but although Heath eventually landed at New York after a long flight in extremely cold conditions, Bill and Eleanor failed to turn up. Floodlights were left on at the airfield and search planes went out, until eventually the couple turned up the next day on a train, having been forced down in New Jersey by the weather and abandoned the plane.[11]

Over in Los Angeles, Jessie was too busy to miss Bill. She'd rented a small apartment close to the YMCA School where she went every day, the only woman in the class, to study radio. With typical Jessie determination, she took a small tapping machine back to the apartment so she could practice Morse code at night. Quickly learning how to repair and maintain a radio, Jessie eventually graduated with high scores that completely satisfied the aircraft company that had funded her study. Only when Kiki had left for England did Jessie fly back to New York. By the time she arrived there, Bill was working as chief test pilot for the company manufacturing the Cirrus engine in America, the same engine that had been in the *Red Rose*. Harry Lyon was now talking about using the plane being built in California for a flight from Los Angeles to Boston and then the UK in the spring of 1930 with Jessie and Bill, but they weren't really taking him seriously any longer and were now making their own plans.

In March, Lancaster was sponsored by the American Cirrus Engine company to make an attempt for the gold medal being offered by the Central Union Trust Company of New York for the first successful round-trip solo light airplane flight from New York south through the West Indies to Georgetown, British Guiana (Guyana). Flying another Avro Avian, powered by a Cirrus Mark III engine, Lancaster would be returning through Panama and Central America, then following the coast of the Gulf of Mexico back to Miami before returning north along the East Coast to New York. It was a total distance of some 10,000 miles that was estimated would take about a month to complete

Five. Burning Bridges

and, presuming that the trip would also prove the Avro was the best light airplane built for that engine, there was an additional $500 prize money. Surely by just a small coincidence, the judges for the competition just happened to include Earhart and Putnam. Leaving New York on 1 March, Bill became the first international pilot to land in Barbados, on the fifth fairway of the Rockley Gold Club, on 29 March.[12]

Jessie, on the other hand, was fulfilling her ambition to at last become a qualified pilot. She'd made this step for more than the obvious reason. Not only would she be qualified to enter races and flying tours, but she was now being approached about speaking engagements in towns to which she'd need to fly, and so being a licensed pilot would give her independence and the ability to go where she wanted at any time. It is possible, too, that she may have heard about the plans for a national women's cross-country race later in the year for which she would need to be a fully licensed pilot. The course for her to qualify as a pilot in New York, where she was based now, was a difficult one requiring her to do at least twelve hours' solo flying time and to pass a written examination about such things as navigation, rules and regulations, rigging and basic mechanics. In those days when airplanes were built of spars and wires, knowing about tension strengths and pulling strains could save your life. With a private license, Jessie would be able to fly any aircraft for pleasure wherever she wished, but it did not permit her to instruct or to carry paying passengers. For that you needed a commercial license, requiring fifty hours solo and more detailed navigational and mechanical knowledge, that permitted you to carry paying passengers within a restricted area. The holy grail of any pilot was the transport license, requiring some 200 hours solo in open cockpit and closed-cabin aircraft and a full knowledge of engines, meteorology, navigation and flying experience cross-country and at night. Jessie flew every day from the training school airfield at Red Bank in New Jersey, practicing cross-country flying and landings in rough air.[13] Even though she was well-known by now, people saw a woman flying so seldom that crowds would gather at the airfield just to watch her, marveling that a woman was at the controls. She flew so often that the airfield personnel knew her by name, as she wrote in a letter home to a friend in Melbourne:

> I have been thoroughly enjoying life lately.... I go out to the field every morning at 8.30 and say, "Hullo everybody, can I have my ship?" They all grin and say, "Sure, Mrs. Miller, we'll get her out." Then I change my clothes and cast an "expert" eye over the bus, climb in and warm up the motor. When I run her up and she gives 1350 revs, I yell, "OK!" and the mechanics pull away the chocks. Then I cast an eye at the wind indicator, adjust my goggles and taxi for the take-off. It's the life! They always call this particular "Waco" plane "Mrs. Miller's ship," as they built an extension on the rudder bar for me as I couldn't reach it. Then I'm banked up with cushions, which always causes amusement.... Am longing to get a plane of my own.[14]

Soon after she'd begun her training, Jessie noticed that her flight time every morning coincided with the schedule of the famous *Blue Comet* train of the Central Railroad of New Jersey as it sped by on its regular run between Atlantic City and New York. As it went through Red Bank, the Comet hit a straight stretch of track and the engineer would usually open up the throttle until the train would near its top speed of 72 miles per hour, about the same speed as Jessie's old training airplane. Jessie began to use the train as a speed marker and after a while this became a good-natured race between her and the *Comet*, although it was never a race she won. The train was inevitably just a few miles an hour faster and would eventually break into the lead as the engineer pulled on the cord, sounding the whistle as the train sped into the distance. Perhaps, after a few weeks of this, the Central Railroad management saw the public relations potential of this race and deliberately encouraged the daily race: according to Anthony Bianculli in his book *Iron Rails in the Garden State*, they eventually paid Jessie to continue it. They certainly arranged a press event on 19 July 1929, for which Jessie was invited to ride in the cab of the *Comet* and finally meet engineer William Smith and fireman Joe Duigan personally. For the occasion, she wore her flying helmet and goggles and presented the men with a bunch of blue cornflowers that matched the color scheme of the train. They in turn gave her a new pair of white coveralls.[15]

Women pilots were so rare in 1929 that Jessie would be only the third to gain her private pilot's license in the State of New York when she passed her test at Roosevelt Field and graduated with License No. 6014 in the first week in April, placing her within a small, select group of women who held any type of pilot's license in the U.S. by that year. The number of women in that group was a fluid figure, increasing during the year as more female pilots qualified. For example, Amelia Earhart obtained her commercial license on 28 March 1929, and so her biographers tend to adopt the female pilot figure of 40 from early in the year. A few months later, however, *Liberty* magazine published a list of the names of the 70 female pilot license holders as of July 29, including private, limited commercial and transport licenses, noting that there were only 70 women out of a total of 6,749 licensed aviators. That article also pointed out that to become a transport pilot cost between $9,000 and $12,000 and two years to qualify, and even then few companies hired women pilots. By December, the figure quoted by the *Women and Aviation* newsletter was 126, including 15 limited commercial and 11 transport licenses.[16] There were even fewer back in Australia. Jessie was only the third Australian woman to be granted a pilot's license anywhere. The first was Hilda McMaugh from New South Wales, who gained her Pilot's Certificate No. 7818 from the Royal Aero Club at the Central Aircraft Company's flying school at Northolt in England in November 1919, followed by Millicent Bryant who obtained her "A" License No. 71 in March 1927, in Australia.

Even while she was training, Jessie continued to talk up her plans for the transatlantic flight at every opportunity. Once qualified as both a pilot and radio operator, she would only need Bill and a navigator on board with her. She had reverted to the original plan of flying from Los Angeles to New York, then non-stop to London, returning to New York and then Los Angeles. If all was still good at that point, they might well fly on to Australia, thus proving the reliability of the trimotor design and providing a chance for much-needed meteorological research. Construction was progressing well on the all-metal trimotor cabin monoplane, Jessie said, which would have a wingspan of 78 feet and a fuel capacity of 1,700 gallons and would be fitted with a pair of 225hp Wright-Whirlwind motors and one Pratt-Whitney-Hornet motor of 550hp, giving it a cruising speed of 100 miles per hour.[17] Then disaster struck.

Still island-hopping to South America, Bill was taking off from Trinidad on his way to Maracay in Venezuela on April 8 when his plane hit an air pocket and smashed into the ground. In a critical condition, Bill was stretchered on board the SS *Vausan*, which was en route to New Jersey. He was landed at Hoboken and taken by ambulance to a hospital in Englewood where he would remain for a long time. They could still dream, but in fact they had lost their window of opportunity for the long-planned transatlantic flight. However, Harry Lyon included Lancaster in plans he announced in May for a 13,500-mile, non-stop, round-the-world flight in September in the trimotor being built, along with Charles Kingsford Smith and Lieutenant Albert Hulse, during which they would refuel in the air. He claimed the flight was now backed for $500,000 at a projected cost of $385,000, but, back in Sydney, Charles Ulm denied Kingsford Smith was involved.[18]

Now that she could, Jessie flew solo out on her small-town summer lecture circuits, using whatever planes were available, but every day she longed for that aircraft of her own. Then during one trip she dived into a lake to cool off and didn't get all the water out of her ears afterwards. Her ear canals eventually became infected and then abscessed and she had to go into the hospital during a tour for minor surgery to enable them to be drained. The next day, she checked herself out and flew on to the next town, despite intense pain and being hardly able to move her jaws. Nothing daunted, she continued lecturing in town after town, for night after night through the early summer while Lancaster languished in hospital, recovering from his injuries. The work on their trimotor had also fallen behind schedule, and so there were not going to be any long-distance flights happening that year. Controversy still reached out to them from back in Australia, too, as they became embroiled in a debate about the fate of the compass from the *Red Rose*. On 31 March 1929, Sir Charles Kingsford Smith and his crew in the *Southern Cross* en route from Sydney to England had made an emergency landing near the mouth of the Glenelg River, in the Kimberley region of northern Western Australia. During the two weeks of searching before they were found alive and well, two of the searchers, mechanic Bob Hitchcock and pilot Keith Anderson in their Westland Widgeon airplane *Kookaburra*, were forced down with mechanical trouble in the Tanami Desert, where they died of thirst. When it was discovered that Hitchcock and Anderson had been using the compass from the *Red Rose*, which may have been inaccurate, there was public and press speculation about the cause of the tragedy. Before his death, Hitchcock had evidently claimed to have seized the compass as payment for money owed him by Jessie and Bill for work he'd carried out on their plane. Hearing of these allegations, an understandably upset Jessie and Bill replied that they had paid Hitchcock a regular salary for being their mechanic and couldn't understand why they were being dragged into this. Because the *Red Rose* was damaged and not in Sydney when they left Australia, they had asked Hitchcock to send them the plane's compass, to which they were sentimentally attached, as it had been their only navigation instrument throughout the England to Australia flight, but they'd never heard from Hitchcock again despite having written to him a number of times. The compass had been entirely accurate when in the *Red Rose*, they pointed out, and the only explanation they could offer was that Hitchcock and Anderson had failed to properly compensate the compass when they refitted it.[19] If a compass was moved from one airplane to another, the magnet in the compass had to be reset to compensate for any change in the amount of metal or even type of metal in the aircraft, and in their haste to leave Hitchcock and Anderson had probably not done this. They were no doubt relieved that the official Committee of Inquiry into the whole affair agreed with them.[20]

Then, as Bill was discharged from the hospital at the beginning of August, Jessie announced that she was interested in establishing a new endurance flight record for women as well as making a solo flight to Europe. Before that, however, she would be taking part in the women's national air race from Santa Monica, California, to Cleveland, Ohio, finishing in time to coincide with the National Air Races events being held there. Amelia Earhart would also be competing and the race promised to be, Jessie accurately foretold, "important and colorful."[21] This would be the first such race for women aviators, and so it would be a unique chance to participate in a historic event that would only happen once. Come what may, Jessie just had to be in Santa Monica.

Six

The Powder Puff Derby

There had been National Air Races in America since 1920, when publisher Ralph Pulitzer adopted a European concept and sponsored the Pulitzer Trophy Race and the Pulitzer Speed Trophy for military airplanes at Roosevelt Field, Long Island, New York, to publicize aviation and of course his newspaper, the *New York World*. For the next few years, the annual national race circulated around various major U.S. cities, primarily promoted as a means of improving aircraft design, to inspire potential new pilots and to stimulate public interest in aviation. These fast and frequently dangerous races involved high-powered aircraft flying cross-country events and field circuits around pylons, and by the 1930s they were attracting huge public attention, featuring in such well-known films as Victor Fleming's 1938 film *Test Pilot*, starring Clark Gable and Spencer Tracey. Famous air racing pilots such as James Doolittle and Wiley Post became household names in similar fashion to modern Grand Prix auto racing drivers.

In 1929 the Nationals were finally brought to Cleveland, Ohio, as an event with enough local civic and industrial support to fully underwrite costs. Equivalent to an aerial Olympic games, this event was held over ten days between 24 August and 2 September. There would be landing contests, flying exhibitions, parachute jumping demonstrations, Goodyear blimp flights and formation flying by the Navy High Hats who flew with their planes tied together from wing strut to wing strut, but what everyone would really come to see were the finishes of the cross-country Derby races, which this year would include for the first time a race for women pilots, as well as the thrilling closed-course pylon races. Danger and death gave these air races an added frisson of fear, much like motor races: six pilots would die during the 1929 events alone.

While up until now major competition air racing had been a male-only sport, in 1929 the Santa Monica Bay District Exchange Club, under the auspices of the National Exchange Club, a men's service club, elected to sponsor the first Women's National Air Derby that would race some 2,700 miles across the country, leaving Santa Monica, California, on Sunday, 18 August, and arriving in Cleveland, Ohio, by 26 August in time for the start of the National Air Races there. Cliff Henderson, organizer of the men's races in Cleveland that year, agreed to manage the Derby with the assistance of an all-male committee. Elisabeth Lippincott McQueen, founder of the Women's International Association of Aeronautics, was also involved in recruiting contestants. The Derby would be patterned after the traditional men's races, two of which would also be taking place about the same time. Though there would undoubtedly be public glory for the first plane across the line at Cleveland, the Derby would be painstakingly timed from point to point in much the same way as a modern car rally, and so the actual winner would be the pilot recording the shortest elapsed time for the entire event. Some $25,000 was at stake in total prize money, but, more important, the competitors hoped that they would gain the fame, respect, and recognition needed in the male-dominated pilot profession to receive offers of work.

The Derby was advertised across America and even internationally, and among the eventual list of twenty contestants were the leading female pilots of that time, including Amelia Earhart, Ruth Elder, Louise Thaden, Phoebe Omlie, Marvel Crosson, Florence "Pancho" Barnes and the only aviatrix from the Southern Hemisphere, Jessie Miller, or Mrs. Keith Miller as she was consistently referred to by the American press. The significance of this Derby to women of the time is hard to understand now in a modern society that is, at least nominally, more gender equalized. This was the first time in any motorized competitive field that women would be recognized as competing on an equal footing to men, and so this unique assembly of leading female pilots was equivalent to a gathering of the leading modern-day Formula One Grand Prix racing drivers. Yet, despite those claims of equal opportunity, it's worth noting that since the inception of the World Championship for Drivers in 1950 only five women have entered at least one Grand Prix and, of those, only two have started in a Grand Prix race. So at the time I write this, although women have gone into space, no woman has become an American president or won a Grand Prix motor race, so there is still room for the kind of female endeavor that took place in 1929.

Jessie really wanted to enter the Women's National Air Derby, but she still didn't have an airplane. Hearing about her plight, a friend introduced her to Lawrence "Larry" Bell, the general manager of Fleet Aircraft, a subsidiary of Reuben H. Fleet's Consolidated Aircraft in Buffalo, New York. When Consolidated's board of directors had been unimpressed by Fleet's new designs for a trainer aircraft for the civilian market rather than military, Fleet had bought the rights to it and set up his own Fleet aircraft manufacturing company, headed by Bell.[1] After talking to Jessie, Bell agreed to supply her with the new Fleet Model 2 after his company had finished some engine tuning and streamlining work to convert the basic trainer and weekend excursion plane into a racing aircraft.[2] The Fleet Model 2 biplane was tough, maneuverable, easy to fly and to land on grass, just under 21 feet long with a wingspan of 28 feet. It was trickier to land on tarmac: the pilot needed to point the nose down the middle of the runway and stay off the brakes. It had a strong, tubular-steel, braced frame and slightly staggered wings, and it was fitted with the Kinner K-5 100hp, five-cylinder engine, giving it a theoretical top speed of around 113 miles per hour. Once the 24-gallon tank was loaded with fuel and the pilot's weight was factored in, though, its cruising speed was more like 85–90 miles per hour, giving it a fuel consumption of about eight gallons an hour if a knowledgeable pilot flew low into headwinds and high with tailwinds. Like the Avro, the Fleet could be easily converted from single to dual control and the tandem cockpit seats were designed to accommodate parachutes, a very useful innovation.[3]

Bell must have been considerably impressed by Jessie's determination, experience and personal charm, because they consented to not only modify the engine but to rebuild the cockpit just to accommodate her smaller frame as well. When work on the Fleet was completed, Jessie went north to Buffalo to collect and test fly the plane at the Bell field in front of their designers and engineers. Flying to Australia was one thing; getting it right in front of a group of highly critical technical experts on a small testing airfield was another. This was their new baby, and if Jessie didn't meet their expectations they'd likely never let her leave with it. However, despite her understandable nervousness, she and the Fleet's performance met the approval of all concerned, and so she finally took off from there to fly solo across the continent to Los Angeles, familiarizing herself with part of the route she would soon fly in the opposite direction.[4]

When she arrived at Clover Field, Santa Monica, on Saturday, 17 August, she found that the office of Jim and Clema Granger, host airport operators for the Derby, was frantically busy as they attempted to cope with contestants continuously arriving and checking in, as well as the race sponsor's constant rule and route changing. Even as late as this, the route was still being altered to include more towns with Exchange Clubs. Adjacent to the Santa Monica Golf Course, which thankfully left fliers a little room for error if anything went awry with takeoffs and landings, and close to the Douglas Aircraft Factory, Clover Field was lined with wooden hangar buildings that often housed the planes of well-known movie personalities and stunt pilots, as well as the large new metal hangar that housed the Grangers' flying school.[5]

While the Derby had brought all these women together who were united in their effort to break through this particular cloud ceiling into a predominantly male environment, they were still competitors who were out to win a race for themselves and for their sponsors. Many women were flying aircraft that had been built or adapted especially for this race, and so their manufacturer sponsors naturally had a vested interest in those airplanes gaining a place. Jessie had already found her Fleet was light and quick to answer the controls, but it was by no means a fast plane compared to what some of the other women were flying. Once fully loaded, the Fleet had a cruising speed of only around 85 miles per hour, perhaps 90 with a tailwind. On the other hand, Marvel Crosson's brand-new single-seat Speed Wing Travel Air with its Wright J-6-7 engine had been clocked at 168 miles per hour. Still, the Derby wasn't just a speed race; it was also about aircraft reliability, flying skills and accurate navigation, and Jessie was confident about her aircraft and her own skills as a pilot. Entrants for the Derby had been divided into two divisions by engine size: Jessie was in the light division for engines of 510 cubic inches or less in the sport planes, and then there was the heavy division for engines up to 800 cubic inches in the larger work planes.

That night at the inaugural banquet, Jessie had her chance to meet her equally confident fellow Derby entrants and to collect her race number. They were a gregarious group of unique women gathered in one place for the first time, and so this Derby became an experience these pilots would treasure for the rest of their lives. For Jessie, from a country on the other side of the world that was still comparatively little known, it was a truly extraordinary experience. Yet of all these women, only Amelia Earhart had traveled by air anywhere near as far as Jessie and even she had been relegated to a passenger in order to be the first woman across the Atlantic in an airplane the year before, whereas Jessie had been both co-pilot and mechanic during her journey between England and Australia. Although this would be the first flying competition for both of them, they already had a lot of experiences to share and quickly became friends.[6] The only other international pilot in the Derby was Thea Rasche, Germany's first female aerobatic pilot, who had become well-known in the U.S. from flying in various air shows. Apparently there were some people who were not entirely happy with women competing in the air, for Rasche had already received an anonymous, enigmatic telegram she had showed to Louise Thaden that simply read, "Beware of sabotage," and suspicions about aircraft sabotage would persist during the event.[7]

As Sunday morning dawned, twenty pilots arrived at Clover Field as the early-morning fog burned off the ground, but there were only nineteen airplanes: a replacement plane for Mary Haizlip had not arrived yet. As the sun rose higher above the waiting aircraft, the Field came alive as last-minute checks and repairs got under way. Like the others, Jessie would have carefully checked that her plane's ailerons were clean, smooth and balanced, because a drooping aileron or dirty surface would add drag as the wind passed over it that would slow the aircraft. She would have checked the wires were tight between the wings and that the rudder control from her pedals was immediate and accurate to left and right, crawled under the plane to scrub oil and dirt off the belly that would also affect speed, and she would have made sure the fuel tanks were full and that the required gallon of water, three days of food rations and a parachute were aboard. Then she would have checked her maps and aeronautical charts on which she would have drawn her flight path, making notes about possible features and hazards and pinpointing the checkpoints and committing it all to memory as much as possible. There wasn't much opportunity to read details on charts while piloting an aircraft like this: long before the automatic pilot was invented, you kept your hands on the controls at all times.

Waiting for the race to start at two in the afternoon, Jessie would have restlessly paced up and down as she typically did, smoking restlessly, probably occasionally exchanging advice and stories with one of the other women pilots or dealing with inane questions from persistent reporters who were unconvinced women either could or should fly airplanes. But these women were pioneers, about to fly in a contest that was the first of its kind, and the newspaper-reading public dearly wanted to know who they were. Charles Lindbergh's flight from New York to Paris in May of 1927, the same year that pilot licensing requirements had been introduced in the U.S., had galvanized the general public into an interest in flying. Within two years, an increasing number of aircraft were being manufactured, more people were obtaining licenses and air racing was all the rage. Military aviators such as Lloyd Stearman and Walter Beech and civilians such as Clyde Cessna were now manufacturing recreational aircraft for civilians who could afford them, and the wide variety of planes on the tarmac that morning reflected the popularity of this new industry. However, there was still a lot of gender bias towards female pilots within an industry where attitudes had not progressed very far beyond 1927 when the manager of the French Air Union stated that "women cannot be trusted with large planes," and the management of the German airline Lufthansa believed that "passengers would not be likely to trust women pilots."[8] Reporters and newspaper editors were often patronizing and condescending when referring to female pilots, using terms such as "the butterfly sex," "petticoat pilots," or "flying flappers" that trivialized women's aptitude as well as this historically significant race. A few days into the race, some Australian journalists even adopted the name "She Derby" for it.[9] As Mike Walker writes about the Derby contestants, "There were many men, perhaps the majority, in the aviation world who were convinced that no one would arrive at all: that far from finding Cleveland, the girls would have trouble finding somewhere to powder their noses."[10] Even the Derby race committee had their limitations when it came to gender: when Opal Kunz first entered her Travel Air with its powerful 300hp Wright Whirlwind engine, they had told her it was "too fast for a woman to fly" and she had to find a plane with a smaller engine. The committee had been unwilling to start the race in California, opting for Omaha, Nebraska, because they didn't think women could fly over the Rocky Mountains until Amelia Earhart had wired them that the race would hardly be taken seriously if the route studiously avoided all danger nor would it be a true "cross-country" race if it started halfway across the country. Then the committee thought that the women would need the help of male mechanics who would travel with them until some pilots heard that Hollywood producers were planning to put starlets aboard planes that would really be piloted by the men and have them met by film crews at stops. Once again, Earhart stepped in, informing the committee that not only was this absurd, but if she and her fellow pilots were not allowed to fly solo, there would be no flying at all. Once again, the committee backed down.[11] The race would depart from California and the mechanics would follow the race in cars. However, the committee won the point that it would be a daylight race only (women couldn't fly in the dark) and the women would have to land at regular rest and refueling stops. In an effort to keep the women grouped together for safety, the last woman to land at the end of the day would be the first to take off the next morning.

Over seventy women had originally applied to take part in the Derby, but the qualifying conditions were strict. Each pilot had to have flown 100 hours solo, 25 of which had to be cross-country hours flown more than forty miles from their airfield. Each pilot had to be licensed by the United States Department of Commerce as well as the Federation Aéronautique Internationale. Each plane had to have an Approved Type Certificate. This list of conditions had reduced the applicants to forty, of whom, for various reasons, only twenty subsequently survived as contestants. Of those, only two were not American: German pilot Thea Rasche and Australian Jessie Miller. In the face of constant insinuation that as women they couldn't possibly do what male pilots could do, these contestants, these airplanes and this race became a significant symbol of women's emancipation and achievement in the air. This was the first time that many of these women had actually met one another, although they would have followed their colleagues' achievements in newspapers, and so for many of the women this was their first experience of being part of a sisterhood of the air. Racing opened the doors to her world, one contestant declared, asking that she not be "cut off from the adventure men have been hoarding for themselves in the guise of protecting me from danger."[12]

By one-thirty that afternoon, bands had played, flying had been demonstrated, Amelia had landed from the Goodyear blimp *Volunteer*, and the contestants had all been introduced. Then Jessie lined up her Fleet biplane on the tarmac with the other eighteen airplanes that waited in two rows for their start signal. An estimated 20,000–30,000 people, among whom could be seen the well-known faces of Will Rogers and Wiley Post, author Edgar Rice Burroughs and cowboy star Hoot Gibson, were there to see them off.[13] As the bright California sun beat down, Jessie would have sat impatiently strapped into her open cockpit, sweating in the heat, checking her instruments and eyeing some of the competing aircraft. Amelia's bright orange five-passenger Lockheed Vega monoplane would be the one to beat, she probably thought, but she wasn't about to readily dismiss Edith Folz's Alexander Bullet, a fast and durable aircraft despite its bad reputation for killing pilots. Both planes were designed with the new enclosed cabins, criticized because pilots wouldn't be able to hear potential engine problems or exit the cockpit quickly if disaster struck. Then Jessie would have focused on the second hand of her watch as it counted off the seconds towards the start. Perhaps she said some words to herself similar to those of Louise Thaden, "This is adventure. May we all come safely through it."[14]

At precisely two o'clock, the sound of a pistol shot that had been relayed by radio from Cleveland resounded from speakers on the field, signaling the starter's flag to drop for the first airplane. Led by Marvel Crosson, nineteen planes were flagged off at one-minute intervals. Jessie, the second to take off, would have headed her plane west into the afternoon breeze from the ocean as she taxied into position, peering around the nose cowling to count down the ten drops of the red flag, feeling herself becoming part of her machine. Then, as the white and red flags fell together, she would have smoothly applied full power, hearing her engine roar in response as she moved off, applying right rudder to compensate for the revolving propeller's torque that pushed the plane left, as she gathered speed down the field. Once the tail lifted slightly, she would have been able to see directly ahead of her towards the spectator crowd as she headed towards them and then, suddenly, she would be up and over them, climbing rapidly as she made the 180-degree turn that would bring her around to the east towards San Bernardino. Down on the ground, media personality Will Rogers shook his head and commented condescendingly, "It looks like a Powder Puff Derby to me," and from then on the Women's National Air Derby of 1929 would be popularly known as the "Powder Puff Derby."[15]

The first leg of sixty-eight miles to San Bernardino was deliberately short because of the late start to allow for the inaugural festivities on that first day. Jessie, along with the other women, focused on following the main road east and tried to ignore the blazing heat of around 100 degrees. Holding the stick between her knees, she would have repeatedly glanced at her aeronautical chart to confirm her route. Her chart indicated terrain variations by color and contour in greater detail than the average road map, much like a nautical chart indicates depth to the seafloor, and it also showed navigational checkpoints such as rivers, lakes, towns and railroad lines as well as compass variation lines. Given the excellent visibility, she would have had no trouble finding her way even though she hadn't actually been over this part of the country before. She'd heard that the local Exchange Club members had even painted "SAN BERNARDINO" in large white letters on the roof of the Fox Court Theatre so the fliers would know for sure it was the right town. But, somehow, Jessie lost her way among the unfamiliar landscape and wandered around for an hour before she finally sighted the plume of dust in the distance that marked where planes ahead of her had landed at San Bernardino's Federal Field. As her propeller crossed the line at the approach end of the runway, an official stationed there signaled the timer who had the official clock and he noted the arrival time before forwarding it to the chief timer. While they

No doubt the San Bernardino photographer had a good reason for using a car as a backdrop for this group of well-known female pilots participating in the 1929 National Women's Air Derby. From left to right, they are: Margaret Perry, Neva Paris, Louise Thaden, Thea Rasche, Edith Folz, Vera Dawn Walker, Ruth Elder, and Jessie. The woman behind the wheel is unknown (courtesy of the International Women's Air and Space Museum, Cleveland, Ohio).

were doing that, Jessie pulled up into a climbing turn and came around to land. Peering through the pall of dust, dodging people and cars and other planes, she finally pulled into a tie-down spot without hitting anything. Opal Kunz, a fellow competitor who was the wife of the vice president of Tiffany's, hadn't been so lucky. The visibility had been so bad for her earlier landing that she wasn't able to gauge her height correctly, and thinking that she was still ten feet above the ground, she had been unprepared when her wheels made contact in a very hard landing, damaging her undercarriage. Still, although she'd only averaged 56 miles per hour for the distance, Jessie hadn't been the last in: Mary von Mach had lost time after being forced down near Montebello, and Amelia Earhart had been forced to return to Clover Field for repairs and then to start again.[16] Even she overshot the runway at Federal, scattering spectators in all directions. "Pancho" Barnes was placed in top position for the day.

That evening, after what would become the customary race-night banquet, the women got together to make the first of their collective decisions. First, they would not consent to Mary Haizlip being disqualified from the race because she hadn't been able to start with them, and they all signed a petition confirming that decision. They then discussed their next stop, Calexico, some 144 miles away. Some of the women who had been there before pointed out that the runway was dangerously short for the heavier aircraft. Yuma was closer and safer, and besides, Calexico had been a last-minute revision to the route and some of them had already made fueling arrangements for Yuma. Jessie wholeheartedly agreed and once again added her name to a petition that declared the women would fly no farther unless routed through Yuma. They got their way. In the early hours of the morning, race organizers once again agreed to a compromise: the women could land at Yuma, but on the way they would fly over the Calexico checkpoint to be identified. Empowered by realizing they had strength in numbers that allowed them to influence such decisions, the women began to discuss forming some sort of national organization for female pilots that would give them a forum for such discussions and perhaps enable them to network jobs.

After only a few hours' sleep, they arrived back at the airfield for their pre-flight checks before taxiing out to wait for the starter's flag. They lined up in reverse order to yesterday's arrivals, which meant that Vera Walker would lead off for the heavy class and Jessie would be first for the light planes. Jessie would have checked her few instruments as she waited, keeping one eye on the starter and one ear on the engine note, sharply attuned to variation in the smooth roar of the pistons. Holding her brakes while carefully teasing the throttle open, she would have achieved just the right amount of power as the flag fell to leap down the field and up into the smooth Monday dawn air towards Calexico. This day's legs would be much longer: 144 miles to

Calexico on the American-Mexican border, then on to Phoenix in about three hours' time.

Visibility in the clear morning air there is excellent and so Jessie would have been able to see various navigation features below her such as roads or rail lines as she looked over the side and then check them with her map as she followed her course drawn on it. Because she was so small, though, she was actually sitting on her parachute and then a large cushion to give her some height so she didn't have to unbuckle her seat belt and pull herself up to see over the edge of the cockpit. As she flew on, it must have struck her that from the air that this countryside bore some similarity to the area around Southern Cross where she was born. The competitors in the smaller planes flew east through Banning Pass and then turned south to follow the eastern flank of the mountains, a longer path that traded the extra distance in order to avoid what would be a long, slow, circling climb over the mountains for them. Making sure they flew directly over the timer at Calexico and were clearly seen, they then leveled out east towards Yuma across a desolate, monotone, sand-dune landscape. The Yuma airfield was a 160-acre square with no defined runway or control tower, only 200 feet above sea level and prone to being obscured by sand, so the women were slightly on edge and made sure they kept looking in all directions for other aircraft as they flew west to east across the field to be timed and then fought their planes around through the turbulent morning thermals to land by the only small building. They climbed out of their cockpits to be assailed by morning heat that Louise Thaden likened to standing in front of an open oven door.[17]

As they landed one by one, news began to circulate among them about their experiences so far, although they could all see what had happened to Amelia's Lockheed. She'd landed too long and her plane had nosed over in the loose sand, bending the prop, so she had to call Los Angeles to have another one flown in. Claire Fahy, a test pilot from the Lockheed Aircraft Company who had married her flying instructor, had been forced to land at Calexico when some of the bracing wires between the wings of her Travel Air biplane had snapped without warning and was claiming she'd been sabotaged. Mary Haizlip, trying desperately to catch up with the others after her late start, overshot the Calexico field in the gathering dusk to enter Mexican airspace and land at the better-lit Mexicali airfield on the other side of the border. Consequently, she'd been delayed several hours filling out explanatory paperwork. Thea Rasche had been forced down in California with a blocked fuel line, which would make her twenty-four hours late. She also suspected foul play. Bobbi Trout had run out of fuel just short of Yuma but on the Mexican side of the border and had crash-landed in a field. By the time her plane was towed to Yuma and repaired, she would be three days behind the others but just as determined as the front-runners to finish. In another consensus decision, those who did land at Yuma's Fly Field that morning all decided to at least wait for Amelia's plane to be fixed before leaving on the next leg.

However, this meant that by the time they were in the air again it was the middle of the day. The temperature was now 120 degrees, and the heat bred thermals that rose up from the desert, creating invisible waves of turbulence that violently buffeted their planes. Even with her seat belt cinched up as tight as she could get it, Jessie's tiny frame would have been lifted almost out of the cockpit as she rode her bucking, lightweight biplane around the sky like a bronco rider in a rodeo. In the absence of any likely landing place in that landscape from Hell, there was no alternative but to ride it out and pray the aircraft had been built on a Monday rather than a Friday. By nightfall, seventeen planes had landed at Sky Harbor Airport at Phoenix, Arizona. Opal Kunz came in late, having wandered off-course. Running out of gas to the north near Prescott, she'd been forced to land in the bed of Granite Creek. Fuel and oil were trucked out to her and she'd only just had enough room when taking off to miss trees. Only Marvel Crosson remained unaccounted for. There were reports of a crash and search parties were called out.

That night Jessie and Amelia Earhart shared a room, as they had at San Bernardino, but this time they had a chance to talk for hours as Amelia gradually lost her reserve. Among many other things, they discussed the public perception that if a male pilot was killed it was all just part of the job, but the death of a female pilot was totally unacceptable. One hears similar discussion in modern times about Western women combat troops serving on front lines. The two agreed that as women's presence in the air grew in the coming years, the public would just have to get used to the fact that the same risks were there whether you were male or female. It was the start of a friendship that would last until Amelia's disappearance a few years later.[18]

The next morning, 20 August, as the search continued for Crosson, the fliers gathered over breakfast to discuss whether the race should even continue. Jessie and Amelia had both advocated that it should when von Mach inspired them all by pointing out that Crosson would not have wanted them to quit now but to carry on and prove the abilities of women pilots and modern aircraft. "Our pain," she declared, "shall become her tribute."[19] They all nodded in silent agreement and then headed for the field where, once again, they went through the litany of refueling and pre-flight checks, while signing autographs and carefully fielding reporters' questions, before lining up their planes for the eight o'clock start.

In third place now in the light aircraft class, Jessie lifted up into the cooler morning air above the palm trees of Phoenix. Their next leg was the 280-mile run farther south to Douglas, Arizona, still on the Mexican border, keeping the Superstition Mountains to the east with the lower Santan Mountains to the west. Jessie and her fellow pilots would be landing at the new, brilliantly illuminated Douglas International Airport that had only been completed at the end of 1928, a huge facility for its time covering some 1,294 acres that actually straddled the Mexican border between Douglas and neighboring Agua Prieta, 654 acres on the American side and 640 on the Mexican. Nevertheless, the race had been going to bypass them until objections by the local Chamber of Commerce influenced organizers to have planes land at Douglas instead of Tucson. So, they followed the railroad southeast to Tucson and flew over the town, keeping sight of the road while wending their way between mountain peaks east to Tombstone. From there, they continued east towards Elfrida, where they would turn south with Swisshelm Mountain to the left to reach the airfield in Sulphur Springs Valley at Douglas. However, as Jessie neared Elfrida her engine began to splutter. To her horror, she was unexpectedly running out of fuel. Frantically finding the nearest flat field before her engine cut out altogether, she landed and then walked a number of miles through sand, sagebrush and cactus before finding a farmer who could spare some fuel for her and give her a ride. When they reached the plane, they found that the pasture was now also occupied by Vera Dawn Walker's aircraft and a very large bull. Vera had strayed off-course dodging bad weather before being forced down by a vicious thunderstorm. Spotting Jessie's plane, she'd figured it was a safe place to land, but by then the bull was objecting to sharing his field, snorting and pawing mud, so she'd decided to stay put. The owner just shooed him away while declaring how friendly and playful he was, and they got down to the business of refueling.[20] Sensibly taking time to inspect her plane thoroughly before taking off, Jessie discovered that the sharp cactus needles had ripped the fuselage fabric. Had she not seen the fine tears, the skin of the plane would simply have shredded once she was in the air and she would have fallen out of the sky. So, after thanking whoever was looking after her, Jessie had to spend hours repairing fabric before she could take off again for Douglas the next day. The delay was a serious blow, but on her arrival, while also muttering suspicions about foul play, Jessie typically declared on landing at Douglas, "While I don't think I have much chance of winning, I am going to keep on trying."[21]

Jessie and Vera hadn't been the only pilots to suffer mishap and misadventure that day, though; apparently Ruth Elder had bull problems of her own. A violent gust of wind shredded her map, according to Thaden and Earhart, and so Elder had landed her plane in a field to try to find out her location only to find it full of curious cattle who liked the taste of the "dope" solution that covered the plane's fuselage fabric. Surrounded by bovines licking her bright red airplane, she was left to pray that there were no bulls among them until she could take off again.[22] However, the worst news by far was that Marvel Crosson's body had been found some distance from her crashed plane in desolate rough country twelve miles north of Wellton, Arizona. She had possibly fallen from her plane when rendered dizzy or unconscious by carbon monoxide fumes, although there have been and still are various theories put forward relating to the cause of her death.[23] Conservative commentators on radio and in the press promptly clamored that air racing was not for women. The *New York American* considered the women's air derby "ill-advised" and that a fine young woman had "been called as a sacrifice on the altar of premature competition. Air racing for women should be discouraged as a far too hazardous adventure."[24] Much more pertinent to the fliers, however, were the claims of sabotage revived by this accident. The press made much of stories that the women's nerves were "on edge" because of the claims, that Earhart refused to allow mechanics to touch her parachute (probably quite rightly, considering what later happened to Blanche Noyes), that another pilot had locked herself in a room, and that Jessie was "near a breakdown" (highly unlikely) and talked of "a jinx" (more likely).[25] Back in Santa Monica, publisher Robert P. Holliday who had been associated with the start of the Derby had sent a telegram to Cliff Henderson, manager of the national races at Cleveland, suggesting that the women's race be halted until all the aircraft had been thoroughly inspected and that the planes had not been given proper protection at the overnight stops. While Henderson promised better security, Floyd J. Logan, the chairman of the national air races, spoke for all the women competitors when he declared the race would not be stopped.[26]

Weaving their way along dry riverbeds and between much higher mountain ranges up to 8,500 feet on their way from Douglas to Columbus, New Mexico, the race contestants took a shortcut through a little corner of Mexico and then picked up a road that took them towards the Columbus airfield. Even in this high, hot, dusty and remote place there was still an excited crowd of spectators, mostly ranchers and miners, waiting to greet the competitors with snacks and cold drinks as they staggered out of their planes under the baking sun to find a spot of shade while the planes were refueled. Then it was back up into the sky and on the way to El Paso, Texas, a straight leg of about seventy miles following the border and the Rio Grande.

As Jessie and the others began to land at El Paso, there were already warnings out for an approaching storm. Survey planes went out and came back with the news they had about two hours before the storm hit, so the women took a vote and decided to tie down and stay put for the night. Planes were quickly walked into hangars where they would be protected and, with that, the race became grounded there for the night as visibility rapidly became non-existent. As the fliers assessed their standings, it was clear that Louise Thaden was well in the lead of the heavy class while Phoebe Omlie was clearly ahead of the light plane group. Bobbie Trout had yet to arrive and Thea Rasche wouldn't reach them until the following day. Margaret Perry was still bravely flying, although she was very ill and running a high temperature.

Meanwhile, back in Los Angeles, the inquiry was still going on as to whether the race had actually been sabotaged. The Deputy District Attorney, Mr. C.O. Thompson, and his colleagues had interviewed eleven witnesses, including San Bernardino airport mechanics, race officials and members of the San Bernardino Exchange Club. Finally, the inquiry decided that the planes had been adequately guarded all night and that no suspicious people had been seen either loitering near or tampering with any aircraft. In short, although sabotage might be suspected, nothing was found to confirm it. Nevertheless, in the future the planes would be kept under guard during the night.[27]

Fortunately, the storm blew itself out and the next morning dawned bright and clear. As the aviators took off from El Paso towards Pecos, the mountains in the distance shone a deep blue, crowned with white clouds. The cooler morning air was smooth flying as they droned over desolate high desert that had its own special beauty, sometimes rising around them, so Jessie was able to quickly reach 4,000 feet, at which she could steer her small plane around mountain ranges, rather than attempting to climb them. Still, the aviators would have had to climb to around 4,000 feet as they flew around mountain ranges rather than over them, and it was hard work during which you had to remain constantly alert. Finally, following a road towards the town as the land fell away beneath them, the pilots began to drop down towards Pecos, which sat at a much lower altitude of around 2,500 feet. When it came into view, just after nine in the morning, the landing field was packed with local townsfolk who had done their best to clear the mesquite and sagebrush from the area to form a narrow landing strip, but then they had parked their vehicles right along the edges of it and were standing around them. They had left no margin for safety at all; any plane that misjudged the strip by even a short distance was going to cause a tragedy.

While Jessie, who was currently placing third among the light planes behind Phoebe Omlie and Edith Foltz, managed to land safely and quickly taxi out of the way of the others, Pancho Barnes wasn't so lucky. At least one car had been driven too far out onto the strip and Barnes, unable to see directly in front of her plane with the nose raised to land, ran right into it. As the dust cloud settled, an irate and swearing Barnes emerged to look back at her wrecked plane. Neither of them would be going anywhere again and she was out of the race, although another Travel Air plane was soon on its way from Wichita so that she'd at least be able to reach Cleveland. Then Blanche Noyes came in behind her with a scorched plane that had a broken wing and crippled undercarriage, bringing it down to land on only the right wheel and sliding to a stop as it settled onto one wing. As her fellow pilots tended to her blackened face and scorched hands, she told them that as she was flying at 3,000 feet about twenty miles out she'd smelled something burning. The next minute, smoke was billowing around her from the baggage compartment behind her shoulders. Finally managing to land in the desert among the thick mesquite, she threw handfuls of sand over the smoldering spare flying suit in the compartment into which, she discovered, a mechanic had dropped a lit cigarette. Once it was out, she had to take off again over ground that was so rough it had damaged her undercarriage. Finding someone at the airfield who could weld it back together, Noyes taped up the holes in the fuselage, braced the wing and took off with the others for Midland, Texas, after they had refueled.[28]

Flying over the flatlands now, where the roads were straight and ran at right angles to each other and forests of oil wells replaced sagebrush, the pilots stopped briefly at Midland and then pushed on towards Abilene, sometimes flying at only 500 feet as they followed the rail tracks that out here were the only significant landmark. The pilots had to follow their course paths on their maps closely and watch their compasses, which the strengthening crosswinds and turbulence during the heat of the afternoon made increasingly difficult to do while flying a plane. You would have to glance at your map, then take a look over the side or move the plane slightly to one side or the other to see ahead around the nose or below you, quickly check the compass and then go back to the map in a constant rotation, all the time jockeying the plane through the sky. It is a constant, wearying sequence of movement but necessary all the same to avoid becoming lost in a landscape with little definition. On the other hand, if in those days you did become "momentarily disoriented" (no pilot ever actually admitted they were lost), you just landed in some farmer's field and asked them where you were. Suddenly, before she realized what was happening, a patch of turbulence at 1,500 feet became a miniature twister that threw Jessie into a "flop" spin, turning her over twice in the air as she scrambled to regain control and tossing her half a mile backwards while she rapidly lost altitude. By the time she got her bearings, the ground was looming fast. Deciding she was too close to the ground to use her parachute anyway, Jessie stayed with the plane and pulled out of her spin only a few hundred feet from the ground. Both shaken and stirred, she determined to finish more than ever now and flew on.[29]

At Abilene, only 138 miles from Fort Worth, the townspeople had not only set out refreshments for the women pilots but had also provided some cots on which they could seize a few moments of much-needed rest. Then it was off to Fort Worth, where twenty thousand spectators controlled by only thirty policemen awaited them. Louise Thaden crossed the finish line first, having flown the leg from Abilene in 5 hours 31 minutes. When Amelia Earhart in her Lockheed Vega landed seven minutes later, the police could no longer control the huge crowd who broke through the lines and surrounded her plane. Gladys O'Donnell was right behind her down onto the grassy field, followed by Ruth Nichols, Phoebe Omlie, Ruth Elder, Mary Haizlip (who had caught up with them all at last), and then Jessie, with the rest of the field touching down safely behind her after the tiring 600-mile flight from El Paso. Edith Foltz was delayed at Abilene while her landing gear was repaired, along with Margaret Perry, who was still very unwell. Blanche Noyes was still waiting at Pecos for her landing gear to be repaired. Now, Fort Worth being a beef town, the fliers were hoping for a change in what had become a perennial evening banquet menu of chicken, but they were disappointed yet again. Chicken it was, hosted by wealthy local *Star* newspaper publisher, oil tycoon and philanthropist Amon G. Carter, who was tricked out for the occasion in silver spurs, leather chaps, ten-gallon hat and a pair of pearl-handled pistols.

At nine the next morning the remaining fliers, except for Edith Foltz, who brought up the rear just after noon, took off on the leg to Tulsa and then Wichita, farewelled by Blanche Noyes, who was delayed again with repairs and had to remain behind. Many of the planes in the race had actually been built in Wichita and the fliers were anticipating a major rest and repair stop because of the facilities. There were fewer of them now, with Pancho Barnes' plane confirmed as too badly damaged to continue and Margaret Perry finally hospitalized in Fort Worth with what was later diagnosed as typhoid fever. Bobbi Trout was still trying to catch up after her Golden Eagle had been towed into Yuma and repaired. As Jessie and the others flew across Texas and into Oklahoma, they navigated by the one-mile-square farming sections below them, maintaining the same angle across section lines to keep them on course. The only geographic feature marking the state border was the meandering Red River. Despite reduced visibility because of a deliberately lit brush fire burning through 25,000 acres of trees and brush, as well as clouds of dust, they all eventually landed safely at Tulsa. Ruth Elder had drifted off-course, while Mary Haizlip had been forced down twice by a dirty oil line and Vera Dawn Walker's engine had overheated. The Tulsa stop proved entertaining when there was an exchange of views about the place of women. In reply to an address by William Martineau, president of the State Press Association, Opal Kunz commented that American women pilots could be involved in national defense in the near future and that she was going to be ready for service. However, self-appointed commentator, Derby critic, and oil-well cementing millionaire Erle P. Halliburton, in his capacity as head of Safeway Airlines, retorted that the Derby contributed nothing to aviation and should be canceled immediately. Women lacked certain qualities that men possessed, he declared, one of which was "handling details essential to safe flying." Amelia Earhart's answer was to invite him into the cabin of her plane, where she expanded his education about women and safe flying for a few lively minutes.[30]

So, after that edifying stopover, the women departed for Wichita early

Friday afternoon. Louise Thaden still led the field and was the first to touch down there at 3:17 in her blue and gold Travel Air, having been accompanied from some distance out by a radio commentator who had met them and was following the women in while broadcasting live. It had been important for Thaden to be first: her family were there to meet her, along with the entire Travel Air factory crew. As Jessie and the other pilots flew over a short time later, they could see below them the large new municipal airport completely carpeted with automobiles and people. Wichita had prepared well for the fliers' arrival, learning from the problems of other hosts along the way. The airfield was solidly turfed and well lit, and a large hangar had been built to house the aircraft, with a mechanic assigned to each aircraft. Each competitor was billeted with a host family, driven around in a car with her name on it, and they were honored guests at the requisite dinner and dance that evening. Journalists were eager for stories about possible sabotage conspiracies, and although Louise Thaden declared, "The rumors of unsportsmanlike color that have been spread about the 'dirty' work done in the race are absolutely false," some individuals revealed suspicion still lurked in pilots' minds. In Thea Rasche's opinion, "there was still some 'dirty work at the crossroads,'" and her forced landing had been due to gasoline that was much dirtier than she would expect under ordinary circumstances. Jessie said she was still wondering why she ran out of gas so unexpectedly outside of Douglas. Although it had been assumed to be her mistake, "it seems like a rather peculiar mistake," she grumbled. Still, the reporter observed, all the women were friends and "none thinks to blame anyone else in the race." As Earhart summed up, "We have worked hard and believe we are accomplishing something worthwhile for aviation."[31] Louise Thaden wrote lyrically that the women felt "hope, determination, a feeling of history in the making with each one playing a part. Adventure, youth soaring carefree on wings of romance, intoxicated, happy, thrilled, suffocated in rapture."[32]

By the time Jessie arrived back at the airport the next morning, the sky was already bustling with aircraft. The racers were not the only people on the way to Cleveland by now; Wichita was a major center of aircraft production and all the newest models were heading north to be exhibited at the trade show that would be taking place in conjunction with the National Air Races. Maneuvering out onto the tarmac, each of the women pilots waited for the flag to drop and then added power as they began to roll, waiting for their speed to build as the tail of the aircraft lifted, waiting for the right moment to take to the air. As they rose into that busy overcast sky, they would have kept a careful lookout for other aircraft as they came around on to their course for the 175-mile leg to Kansas City, from where it would be another 250 miles to East St. Louis. Kansas City was only a quick stop to refuel and grab a bite to eat. While most of the women ate a quick lunch from tables that had been set out in a hangar, some of the really eager pilots didn't even climb down from their planes. However, Neva Paris, happy to have arrived in her hometown, just had time to run from plane to plane to let them know of a planned meeting about forming a pilots' association that would take place under the grandstands after they all arrived at Cleveland. Typically, the reporters for the local Kansas newspapers were fascinated by what the women were wearing as well as Jessie's unique Australian accent that they had never heard before. They noted that while Jessie was light as a jockey, Phoebe Omlie was heavier and sunburnt and wore a cloth hat, Gladys O'Donnell wore white coveralls, von Mach a black shirt and cord breeches, Amelia was freckled, tanned, and looked tired, and Ruth Elder was now flying to her wedding, having decided on the way to accept Walter Champ's proposal.[33]

The fashion review over, they flew on to St. Louis, following the Missouri River though increasingly hazy conditions to land at the Parks Air College field mid-afternoon. The flying school college had only been established the year before on the eastern side of the Mississippi River by Oliver Lafayette Parks, a former Chevrolet salesman who had seen a better future in selling planes and flying lessons. He was so successful that before the end of 1928 his landing field was better equipped than St. Louis' Lambert Field and his 400 students made this the largest flight school in the U.S. Still, it wasn't a big landing area by today's standards, nor were the perimeters obstacle-free. By the time they avoided trees and power lines, some pilots found they were landing halfway down the field and rapidly closing on the perimeter fence. Blanche Noyes, the first in, damaged her landing gear and Neva Paris ended up nosing onto the perimeter road with her tail in the air. Jessie, possibly the most experienced at landing her plane anywhere at all, made it down safely. Thea Rasche and Mary Haizlip were still behind them somewhere contending with fuel line problems, and Bobbi Trout was about a day behind them back at Wichita but still intending to flying on.

Dense fog on the ground delayed the start the next morning, so photographers took the opportunity to snap some group shots of the fliers while they could. The Cleveland Aeronautical Exposition was opening that day and the Derby was scheduled to end there on Monday. Some 100,000 people were expected to be there for the opening day and by the time the Derby was due to finish there would be considerably more. Three men's cross-country Derbies were also due to finish in Cleveland: one from Portland, Oregon; another from Miami Beach, Florida; and another from Oakland, California. Louise Thaden was comfortably in the overall lead in the women's race; with only 536 miles to go it was unlikely, bar accident or forced landing, that anyone would now overtake her. Phoebe Omlie was leading in the light plane class. Jessie, who had not been able to make up that long period she'd been on the ground in search of fuel, was currently holding third place. The airplanes

Six. The Powder Puff Derby

A group photograph of the Derby pilots taken at the old Parks Field, also known as Parks Airport, at St. Louis while waiting for the fog to clear in August 1929. They are, from left to right, Mary von Mach, Jessie Miller, Gladys O'Donnell, Thea Rasche, Phoebe Omlie, Louise Thaden, Amelia Earhart, Blanche Noyes, Ruth Elder, and Vera Walker (courtesy of the International Women's Air and Space Museum, Cleveland, Ohio).

had undergone thorough checks overnight during which Louise Thaden's mechanic had discovered her magneto points appeared to have been tampered with. He dutifully insisted on sleeping with the plane that night.

Once the fog burned off, the women were soon in the air, following the road and railroad tracks towards Terre Haute, Indiana, where they once more dodged a poorly controlled crowd on the field to land and refuel. Ruth Elder's heart missed a beat when her engine went dead as she crossed the finish line, so throwing her plane into a sharp banking turn, she managed to make a safe "dead-stick" landing. There was another brief stop for the competitors in the afternoon at Lunken Airport, Cincinnati, and then it was on to Columbus, but after only about thirty minutes in the air Jessie's engine began to cough and cut out. She was flying at about 1,000 feet and wisely figured she'd better land while she still had some power. "I saw this nice field with a lot of grass,"

Jessie in the cockpit of her Fleet biplane at the old Parks Field, in August 1929, during the Women's National Air Derby (used with permission of the Saint Louis University Libraries: Archives and Records Management, Henry Schnittger Photo Collection PHO 63.0.1).

she told reporters later, "and I just 'sat down' in it. I missed some horses, cows and pigs, and skipped a couple of ditches, but neither the plane nor I was damaged." She had landed a short distance west of Xenia, Ohio, on the edge of a large pit in a field on the Shoup farm just after five that Sunday afternoon, discovering afterwards that her compass was faulty and so she was slightly off-course. After locating the nearest telephone, she called the Columbus airport, from where some mechanics eventually arrived. The engine had lost compression in two cylinders; perhaps the valves were sticking, but the problem couldn't be fixed that night and so Jessie sent word that she'd join the race in Cleveland the next day and retired to a local Xenia hotel. Another pair of mechanics flew in early Monday morning, and as a result of their combined efforts Jessie was able to be back in the air just after nine o'clock.[34] Bobbi Trout was struggling, too. Having nearly caught up with the group after some amazing flying, she'd also been forced down again with engine problems just out of Cincinnati, but she was able to be off again before long.

Landing in Columbus carefully because of new runway construction, the racers congratulated one another, hardly able to believe they had made it this far, with only one night to go. Some of them felt sad that this unique camaraderie could soon be over and vowed to make sure that it wouldn't end at Cleveland. As Louise Thaden said that night, "We helped each other, worried together, laughed over mistakes, silently wept and endured in community, recognized our strengths, and combated weaknesses. We never mentioned the afraid times."[35]

As Day 9 of the race dawned, Monday, 26 August, this last day of the race saw the fliers once again carrying out their pre-flight checks, talking to reporters and signing autographs. Some took off for test flights, including Ruth Nichols, holder of only the second transport license

Jessie reading a telegram at Parks Field during the Women's National Air Derby (courtesy of the International Women's Air and Space Museum, Cleveland, Ohio).

(enabling her to carry passengers) issued to a woman. Almost down on the ground as she came in to land, she was hit by a crosswind and slammed into a steamroller parked on the edge of the runway, somersaulting to land upside down. As the women ran towards her, she crawled out unhurt, but her third-place standing was lying there in the wreckage, to be picked up by Amelia Earhart. For the only time during the race, the female fliers lined up side by side and took off from Columbus at one-minute intervals in the order of their actual race standing so that they would land at Cleveland in the same order. Consequently, Louise Thaden led off, followed by Gladys O'Donnell, then Amelia, then Blanche Noyes and so on.

Cleveland was only 120 miles away. This year the Nationals had drawn such huge crowds that the bar had been raised for future annual shows. The

The contestants' airplanes lined up for takeoff at Parks Field on the morning of the eighth day of the Women's National Derby, 25 August 1929. Jessie's Fleet, No. 43, is third from the right, and if you look very carefully you can just see her over the top of the "3" talking to a group of people. In the foreground is Louise Thaden's Travel Air, and next to it is Gladys O'Donnell's *Miss Long Beach*. Beyond Jessie can be seen Blanche Noyes' plane and then the Alexander Bullet of Edith Folz (used with permission of the Saint Louis University Libraries: Archives and Records Management, Henry Schnittger Photo Collection PHO 63.0.3–8).

inaugural parade down Euclid Avenue that opened the festivities, consisting of 200 floats, 21 bands and 1,500 marchers, was watched by a crowd estimated at 300,000 while three Goodyear blimps flew overhead. Over 100,000 spectators were there just for the opening day! Thaden crossed the white chalk finish line at 170 miles per hour in a powered descent that included a sideslip and a fishtail that thrilled the crowd. Thousands of people rose to their feet and cheered as she kissed the runway in a slow, neat three-point landing after flying just over 2,700 miles over nine days, in 20 hours, 19 minutes and 4 seconds of elapsed time. Knowing that others such as Jessie and Bobbi were still a long way behind her, she graciously commented to waiting reporters and news cameraman, "Well, folks, the sunburn derby's over. I happen to come in first not because I'm a better pilot than any of the rest of the girls but because I have a fast airplane and I had good breaks. They're all good women and

Six. The Powder Puff Derby

everyone who flew in the race should have first place.... Everybody's been lovely and I hope they're just as nice to the rest of them. They all deserve it." Later, Louise would declare that she had won the cup for Marvel Crosson. Jessie eventually made it into Cleveland and was grinning from ear to ear as she was awarded a third place in the light aircraft division, right after Phoebe Omlie. The place didn't really matter, though; the only entrant from the other half of the world had made the distance. In the words of the humble *Dubbo Liberal and Macquarie Advocate* back home: "Bravo Mrs. Miller! Can't Keep a Good Girl Down.... Well done, Aussie!"[36]

However, Jessie wasn't allowed to rest for long. On landing, she discovered that Fleet Aircraft had entered her in the fifty-mile, closed-circuit speed pylon race for women. It was five laps of a ten-mile course. The bets were in that Phoebe Omlie, Amelia Earhart or Lady Heath from England would easily beat the newcomer, but the little Australian had a surprise for them all. Despite having to hurriedly find a replacement plane when the wheels of her Fleet struck a soft spot in the field while being taxied by the mechanic, wrecking it, Jessie initially came in second, having been timed at some 99 miles per hour. Then Phoebe Omlie, who had been slightly faster, was disqualified for crowding the plane of another flier while banking in a turn around a pylon, and Jessie was declared the winner, with a prize of $1,700. Jessie then won third place in the Women's Australian Pursuit Handicap Race after Thea Rasche and Louise Thaden. The Women's National Air Derby was a landmark event in changing attitudes towards women pilots for all time. After it was over, each of the participants received a bracelet bearing the winged badge of the National Air Races along with a letter from the managing director, Cliff

Jessie speaking to interviewers after winning her Cleveland National Air Races event in 1929 (author's collection).

Commemorative envelope of the National Air Races, dated August 24, 1929, and signed by Jessie (author's collection).

Henderson, in which he commented on the Derby's significance. "If ever there was a question as to women's ability to fly and to take a significant part in this great industry," he wrote, "it is now definitely and finally settled, and their important role in future development will be an accepted fact."[37]

Towards the end of the Cleveland races, Jessie, Amelia, Phoebe Omlie and many of the other women pilots met under the grandstands to talk about forming a professional women pilots' organization. Somewhere about that same time, according to pilot Margery Brown, a man at an airfield had commented to her that women pilots ought to organize. The more she thought about that, the more she thought it a good idea, and so she, Neva Paris, Frances Harrell, and Fay Gillis sent out a letter to all 117 licensed women pilots in the U.S. inviting them to a meeting in November at Curtiss Field, Valley Stream, at which they could discuss forming an organization for women pilots within which women could maintain friendship and business networks and that would provide them with a unified voice within the aviation industry. To her surprise, Phoebe wrote back and told her about the Cleveland meeting at which they, too, had discussed organizing.[38] So on 2 November 1929, Jessie and twenty-five other licensed women pilots flew, drove, caught a train or walked to Curtiss Field, where they met in a hangar. Tea and cookies were served from a spare parts wagon while their organization was discussed over the din of mechanics working on a Wright Whirlwind engine. They discarded such names for their organization as Gad Flies, Climbing Vines, Bird Women and Skylarks before Amelia Earhart, who would become their first president, suggested naming it after the number of charter members, which was eventually settled at ninety-nine, and it has been known ever since as the Ninety-Nines. The very first national organization of women pilots in the world, it is now an international organization of women pilots, divided within the U.S. into eight geographic sections subdivided into some seventy chapters; only two countries outside the U.S. have warranted sections: Canada and Australia. The famous Australian pilot Nancy Bird Walton became the third Australian member of the Ninety-Nines in 1938, followed a few years later by Lady Maie Casey and subsequently many other Australian women pilots, all following the route first flown by Jessie Miller. So as one decade ended and a new one began, female aviators were "becoming inspired, as a class, with the desire to help the other woman," as Margery Brown wrote in 1930. "Not everyone for herself, but each one for all."[39]

SEVEN

Flying the Ford Tour

No sooner had Jessie finished one national race than she was immediately plunged into another. The Fairchild Aircraft Company had been so impressed with her flying in the Derby that in late September they offered to sponsor her for the Ford National Air Tour, for which she would be piloting a gleaming white Fairchild KR-34 biplane. Aviation entrepreneur Sherman Mills Fairchild had recently acquired a controlling interest in the Kreidner-Reisner Aircraft Company based in Hagerstown, Maryland, and the KR-34 was a new development of their already popular Challenger C-4 aircraft. The KR-34 was fitted with a 165hp Wright J-6 Whirlwind five-cylinder, radial piston engine, giving the plane a top speed of 120 miles per hour with a cruising speed of around 90 with a range of some 500 miles at a ceiling of 14,000 feet. It had two open cockpits within its welded steel fuselage that was 23 feet long with a wingspan of 30 feet. It featured an entrance door to the front cockpit and a baggage compartment behind the rear cockpit accessible by a door on the left-hand side. A durable, general-purpose aircraft that could be fitted with skis and floats as well as wheels, the KR-34 would be ideal for use as an air taxi and airmail carrier and as a bush plane in remote locations such as Alaska and northern Canada. As a public relations venture, Fairchild Aircraft planned to enter at least three of their aircraft in the Tour: painted all-white, they would form the Fairchild White Fleet, as part of which Jessie dressed to match in white kid-leather jodhpurs and flying jacket lined with lamb's wool, a white silk shirt with a black tie and a matching white leather helmet.[1] This fashion statement was a long way from the shorts and open-neck shirts she'd worn in Australia and demonstrated Jessie's increasing sophistication and her developing awareness of her public image as a career aviatrix. Only a month later, Elsa Schiaparelli would show a collection in New York that would include an aviation suit, and in March 1930 the *Ladies Home Journal* would publish her article, "Smartness Aloft," that included a photograph of her pilot's costume design, a natural-colored linen crash overall with long trousers fitted tightly at the ankles, patent-leather cuffs, zipper fastenings, a black knit hat

Jessie being presented with her Fairchild KR-34 by Sherman Fairchild (right), with Carl Reed (left), at the Fairchild Aviation Center and Flying Field, now Republic Airport, East Farmingdale, Long Island (used with permission of Long Island–Republic Airport Historical Society).

and collarette and low-heeled shoes. Amelia Earhart released her own line of aviation wear in 1933 that she claimed used actual materials such as parachute cord ties, ball-bearing belt buckles and wing bolts and nuts for buttons.[2]

This 1929 National Air Tour was the fifth in an annual series. When first organized in 1925, it was known as the Commercial Airplane Reliability Tour for the Edsel B. Ford Trophy, but this unwieldy mouthful was quickly abbreviated to the Ford Reliability Tour and later to the National Air Tour or simply Ford Tour. The Ford name is usually associated with cars, but Henry and Edsel Ford were also quite involved with aircraft innovation and manufacture: the Ford trimotor was the first all-metal transport plane manufactured in America.[3] Ford engineers developed the first radio guidance beacons and installed the first concrete runways in the world between 1928 and 1929 at the 360-acre Ford Airport in Dearborn, Michigan. Reliability tours for airplanes were modeled on the reliability tours for automobiles, which had been started in 1904 by Charles Glidden, basically to promote automobiles as a means of transport, and in 1925 a group of Detroit businessmen agreed to a suggestion by Harvey Campbell of the Detroit Board of Commerce to plan a Glidden Tour of the Air. However, when Edsel Ford donated a magnificent four-foot-high gold and silver trophy worth some $5,000 for the Tour, his name became attached to the Tour as well as the prize.

Like the earlier Derby, these Reliability Tours were not speed races but were instead promoted for their competition fairness over distance. This was not a competition for military aircraft but only for civilian planes and pilots. Much as in cross-country car rallies, pilots and aircraft were handicapped and judged on completing stages between checkpoints accurately, in good mechanical condition and on time with fuel efficiency. Each plane had to be capable of 80 miles an hour with a full contest weight load. To that end, competitors were scored using a rather complicated points system based on a "figure of merit" determined by tests carried out on each plane before the Tour began. These tests were in five categories in which you were scored higher according to the greater load the plane was certified to carry, the greater speed of the plane fully loaded, the shortest time/distance in which the plane could take off fully loaded (or "unstick") after the wheels began rolling, the shortest time and thus distance in which the plane could come to a dead stop (or "stick") after the wheels touched the ground, and the least engine capacity (cubic inches) compared to number of cylinders. As a pilot passed through each checkpoint along the 32 stages of the Tour route, provided they completed each stage within the allotted time, their "figure of merit" was added to the stage score.[4]

Even in 1929, none of the competing planes had retractable landing gear or modern wing flaps. Superchargers and variable pitch propellers were not in commercial use, the planes were not fitted with radios, and only some would have had instrumentation like turn and bank or airspeed indicators. The first Tour in 1925 visited just twelve cities around the Midwest and returned to Detroit within six days, but each subsequent Tour had covered more stops and so they had become longer as their popularity increased. Covering over 5,000 miles and 32 city and town stops through Canada and the U.S. with 29 entrants and 17 support planes, this would be the biggest Reliability Tour yet. The field of aircraft varied widely, from a tiny Moth biplane to the huge Curtiss Condor that carried eighteen passengers and a crew of three.[5] When a Tour landed at a town, crowds of people would turn out to see the latest aircraft designs and the top pilots flying them, and often a new municipal airport or landing field that had been constructed for the occasion. In this way, the Tours introduced millions of people to the viability of reliable commercial air travel and were one of the most successful promotional efforts of the early twentieth century. This particular Tour was scheduled to finish on 21 October, the fiftieth anniversary of Thomas Edison's invention of the incandescent electric light, for which Henry Ford was planning a major celebration in which this Tour would play a significant role.

As morning dawned over the Ford Airport at Dearborn on Saturday, 5 October, it revealed a scene that at first glance seemed chaotic. Torrential rain and storms had delayed Jessie and the other pilots arriving from the East Coast who had been attempting to cross the Pennsylvania mountains from New York for five days and were consequently late registering. Bands were playing, engines were warming up, spectators were milling about and photographers were pushing and shoving to get the best shot of the three "sweet lady bird pilots," as they were known: Jessie in her Fairchild, Mary Haizlip from Kansas City in her American Eagle biplane, and Frances Harrell from Texas piloting a Gipsy Moth. Judges scrambled to check correct weights and pilots yelled at mechanics who yelled back.[6] Then suddenly everyone was running for the barriers as the starter stepped up with his flag. The twenty-seven aircraft would leave at one-minute intervals. Always two hours ahead of them would be chief Tour referee Captain Frank Hawks, who would be the first to arrive at a control point to make sure everything was organized for their reception and timekeepers were in place.[7] Following them, in what became known collectively as the "aerovan," were a number of support planes carrying service personnel and officials, including a flying weather bureau managed by government meteorologist C.G. Andrus, a Firestone Ford trimotor official press plane, the *Detroit News* plane, the Wright Aeronautical Corporation's Ford trimotor service plane known as the "flying garage," and the Pratt & Whitney Vought Corsair service plane. Bringing up the rear was the last man to leave every control point: E.W. Cleveland, the chief starter.[8]

As the flag fell for the first time at ten o'clock precisely, C.W. Myers of Cleveland was the first to take off in his red trainer biplane. Their first stop was only fifteen miles away over the Canadian border at Windsor, Ontario. Such a short distance barely gave some of the larger planes time to warm the engines up and gain altitude, but it wouldn't have hampered Jessie in the smaller Fairchild and they were landing in Windsor after only a few minutes. After lunch, the Tour took off for Toronto, which would be their first night stopover. As would be the case at all midday stops from now on, the planes took off in order of arrival. For morning takeoffs, after night stopovers, the order was reversed so that the last to arrive took off first. This method actually kept the planes bunched up better so they looked more spectacular when flying over towns and landing at airfields. As usual, reporters at Toronto were still more interested in what Jessie was wearing than her flying. With one noting that she jumped out of the plane "as gracefully and nonchalantly as if she had not piloted a plane all the way from Detroit in an hour and three-quarters but were stepping out of her car for matinee," her flying outfit obviously made such an impression on the reporters that their first question was an astounding, "Do you think women will ever regard aviation more seriously than an opportunity to wear a smart costume?" One can only assume that Jessie must have by now been quite resigned to such inane questions, because instead of dropping a wrench on the reporter's foot she just took it in her stride. "Why, of course," she replied politely, perhaps through clenched teeth. "Why should one not dress appropriately. Pants, for instance, are best; just imagine how skirts catch and get in the way, besides pants are warmer." She reminded them that of course she didn't wear this kind of outfit all the time and "returned to normal" when back on the ground. Nevertheless, as if Jessie as an independent, achieving woman had been on trial for not being a mother, the journalist's final judgment was, "She has no children. She admitted domestic accomplishments did not interest her."[9]

After an early Sunday morning breakfast at five-thirty, the Tour took off from Toronto for the mid-day stop at Ottawa, some 228 miles to the east over some rough wilderness terrain, not the sort of country in which you'd like to be forced down. They arrived ahead of schedule courtesy of the tailwind. On landing, Mary Haizlip revealed she'd had two fingers broken by some object striking the plane in the air while she had her hand resting on the front of the cockpit, but she'd kept right on flying through the pain. After a quick lunch stop, during which Haizlip's fingers were bandaged and splinted, it was on to Montreal, where the fleet of over forty competing and support planes landed between three and four in the afternoon of Sunday, 6 October, just before the weather broke. The city had in fact incorporated the Tour into the First Canadian Air Pageant celebrations over that weekend, sponsored by the Montreal Light Aeroplane Club, that included races, aerobatic demonstrations and landing contests, and a formation flying team of nine planes from the United States Army Air Service. John H. Livingston was already in the lead on points, a position he'd maintain throughout the race.[10] It rained all night, but getting to their hotel through the storm was the least of the pilots' problems. Over 40,000 people had gathered at St. Hubert Airdrome to meet the Tour and hundreds more clogged the wet streets. It might have taken Jessie and her fellow fliers about an hour to fly from Ottawa to Montreal, but it took the quickest of them over three hours to get from the airfield to their hotel, and most of them took longer than that. The reception banquet scheduled for the early evening at the Mt. Royal Hotel had to be postponed until late that night.[11]

On Monday morning it was back into the U.S., flying to Portland, Maine. Although the White Mountains beneath them would have been clad in brilliant Fall colors, Jessie and the other pilots could hardly see in front of them. The farther they flew, the worse the weather became. As snow, fog and rain combined with forest-fire smoke, their visibility range deteriorated alarmingly. Unable to fly above the weather and risk losing sight of landmarks, they were left with little option but to fly down the valleys below the cloud in order to keep ground features in sight by which they could navigate. Landing at Portland in the rain, the Tour competitors were welcomed by a dedicated group of Customs officials who descended on the planes to carry out some detailed inspection for any onboard contraband alcohol from non–Prohibition Canada. Some of the fliers took note that the official press plane failed to turn up. Taking off again, they battled headwinds all the way to the overnight stop at Springfield, where they were greeted by some 70,000 spectators as they landed at around four in the afternoon just as the weather cleared. It was a well-organized stop, with official cars meeting each plane after they had taxied into line. After the customary evening banquet, they all watched film that had been shot of the planes coming in that afternoon.[12] Then, on Tuesday morning, 8 October, it was a 100-mile leg to Roosevelt Field at Long Island and then it was another short hop over water against a strong headwind to the new Camden Central Airport in New Jersey, at that time the landing field for the Philadelphia area, where Wiley Post led the field down.

On Wednesday morning, the Tour set out for Baltimore's Logan Field in such fine weather that Jessie and every other competitor achieved a perfect score for the stage, the first time that had happened.[13] Just as Jessie was touching down that afternoon on the narrow L-shaped Byrd Field at Richmond, Virginia, the huge crowd broke through the barriers onto the runway. Suddenly, without warning, a small boy ran out in front of her plane. Desperate to avoid a tragedy, she sharply turned the airplane to one side, putting too much strain on the undercarriage. As the wheel strut snapped, the plane sagged over and the wingtip scraped the ground as she slid to a stop. Shaken by the near miss, she was congratulated for her quick thinking by Tour officials who refused to deduct any demerit points for the landing and made sure her plane was repaired overnight. However, she wasn't the only pilot who had a problem with that crowd; Roger Williams, piloting one of the official planes, narrowly missed plowing into an entire group of overenthusiastic spectators running onto the field.[14]

At ten on Thursday morning, the Tour flew out of Richmond towards Miller Municipal Airport at Winston-Salem in North Carolina, from where, after lunch at the Hotel Robert E. Lee, they departed for Greenville in South Carolina, the twelfth leg of the Tour. Mary Haizlip wasn't with them, stranded back in Richmond waiting for a replacement engine part. With few major landmarks beneath them Jessie, who was now in tenth place on the points score ladder, and her fellow pilots had to watch their compasses closely. They touched down in the middle of the afternoon before another huge crowd at Greenville, where, after they had lined the planes up on the airfield for public inspection, they were hosted at a dinner and dance that night. The following day they flew through the warm, balmy air of the South towards Augusta, but during the afternoon the weather deteriorated quickly as they neared Jacksonville, just across the Florida border. For miles in front of them, the pilots could see nothing but overflowing rivers and flooded fields. A stiff headwind meant that they had to fly low in order to maintain sufficient speed to meet the tight stage schedule, and it required a lot of concentration. At that height, they could see the faces of people stranded on rooftops along with their chickens and hogs and could count the cattle stranded on small islands. The pilots were so low under the cloud there was no room for correction in the event of engine trouble, so it was with deep sighs of relief that they all climbed out of their cockpits safely at the end of the day, thanking their engine maintenance crews and modern engine designers that there had been no mishaps.[15]

On Saturday, 12 October, the field turned north in much better weather towards Macon, Georgia, but on the way to Candler Field at Atlanta in the afternoon the advance plane broke down and there was no warning to Atlanta of their arrival. The whole fleet swooped down out of the sky before they were expected, causing some confusion. For the first time in Georgia that anyone could recall, a bridge of aircraft extended at one-minute intervals from one stop to another: at about the time the leader arrived in Atlanta, the last plane was taking off from Macon. At this halfway point of the Tour, according to the number of legs, John Livingston was leading in his WACO biplane followed by Arthur J. Davis, also in a WACO, and then M. Zeller in a Ford trimotor. Jessie was in ninth place, while Frances Harrell was in twenty-fourth place with Mary Haizlip behind her in twenty-fifth. For the first time on the Tour, Jessie, Frances and Mary were delighted to be officially met by a group of women, the female members of the aviation committee of the local Chamber of Commerce. Jessie and the others expected to catch up on some much-needed rest during their weekend layover until Monday morning, but for once the Tour had some competition. An important football game was scheduled in Atlanta that weekend as well, and the fliers' hotels had been overwhelmed. Pilots ended up in rooms scattered all over town, sharing them with the ground crews. One of Jessie's companion female pilots discovered she was sharing a room with a male mechanic. When Jessie went out to inspect her plane on Sunday afternoon, along with some of the other pilots, it was nearly impossible to get onto the field because of spectators and cars. Finally managing to reach their planes, they found people writing graffiti on them, striking matches on the outer skin and stealing souvenirs. They couldn't stand to watch and went back to their hotel rooms, praying there would be enough left of the planes to fly on Monday morning. Sounding very much like the Women's Derby pilots, Livingston would later write that "adequate protection should be furnished at every stop on these Tours, instead of the almost total absence of guards experienced at Atlanta."[16]

To the relief of all, when they inspected their planes Monday morning nothing substantial was missing and they all rose into the air towards Murfreesboro in Tennessee, some 191 miles away. For this second half of the Tour, most of the legs would be longer than they had yet been. Once again, there was a large crowd to meet them and the pilots were entertained to lunch on the roof of the clubhouse. While they were waiting for their food, a fight began down on the field over who was supplying gas to the planes. While Shell had the supply rights to the field, some of the planes were being supplied by Texaco and Standard Oil, but their tankers weren't being allowed onto the field despite Shell having only two tankers to supply forty planes. Needless to say, time ran out to fill them before they were to be waved off. It meant that some pilots lost time through no fault of their own, about which they remained unhappy all the way to Lunken Airport at Cincinnati, where, ironically, Frances Harrell crashed out of the Tour when her plane ran out of gas.

When Jessie and her colleagues arrived back at Lunken Airport early Tuesday morning, a chilly fog had enveloped the landing field, but it burned off just before the scheduled ten o'clock start and they enjoyed fine weather all the way to Louisville and then on to St. Louis. On Wednesday the Tour moved on to Springfield, Missouri, only to encounter what they came to call the Springfield jinx. From now on, they unanimously decided, all towns named Springfield would be excluded from the Tour.[17] With few landmarks visible again, they had to fly through extremely rough air relying on their compasses alone, enduring air pockets that threw them around violently, but having touched down here only a few weeks before with the Derby, Jessie would have at least known her way. By comparison, and as if to prove the jinx, the afternoon leg to the new airport at Wichita was absolutely perfect. Because Wichita was, of course, the home of the Cessna aircraft, the plan had been changed to allow them to take off first from Springfield. Nevertheless, even though Wiley Post took off in his Lockheed five minutes after Lacey in the fastest Cessna, they ended up racing neck and neck into Wichita.

Mary Haizlip led the Tour planes into the air from Wichita as they flew on to St. Joseph, Missouri, in balmy weather and a tailwind, where more than 3,000 spectators lined the field as the planes began to land about eleven on Thursday morning. The first down was Wiley Post in his Lockheed Vega. Once the aviators had been fed at long tables set up in the largest hangar, they took off again for Des Moines, where they had another enthusiastic reception. The following Friday noon they were at Cedar Rapids, from where they beat bad weather into St. Paul, despite some careless spectator setting the grass on fire at Cedar Rapids. Spectators fought the fire sweeping towards the line of planes before it was at last put out. On their way to the Alexander

Airport at Wassau, Wisconsin, on Saturday morning, however, the cloud layer settled lower the farther they flew and soon Jessie and the field were skimming treetops, but they finally found their way in. Then it was on to Milwaukee and an enjoyably social Saturday night. They left Milwaukee for Moline on the Sunday morning in a slight drizzle but flew out of it, turning east in the afternoon for their last overnight stop that night, timed for the dedication ceremony of the brand-new Curtiss Airport at Glenview, Chicago, on 20 October and the big air show there.

Although Chicago Metropolitan Airfield had been established in 1927, the already industrially polluted air over the city made flying a challenge. So, the Curtiss Flying Service, founded by American air pioneer Glenn Curtiss as a subsidiary of his Curtiss Airplane and Motor Company, moved to the more rural area of Glenview outside Chicago to build their new state-of-the-art airport that they hoped would become the main Chicago air travel hub. A one billion candlepower beacon light was installed at the north end of the field to enable flight operations to continue after dark. The centerpiece of the field was Hangar One, the combined hangar and terminal building that was said to be the largest in the world at the time, included in which were a number of innovations such as glassed-in public galleries for visitors to watch mechanics working on the lower floors, or they could relax in upper-floor restaurants and lounges while waiting for flights to be called over loudspeaker systems. The day-long air show dedication of the Field was one of the most impressive occasions Jessie had ever seen. All the notable Chicago business names were on the program: Armour, DuPont, McCormick, Vanderbilt, Walgreen and Wrigley. Some 35,000 people crowded the area, while more than a hundred planes passed in review as part of the Pageant of Aerial Progress. Jessie would have been impressed to see seven parachute jumpers bail out of a Ford trimotor at the one time, and the day seemed to be over much too soon.

In the cold dawn of 21 October, the pilots prepared to leave a slightly quieter Curtiss Airport on the last day of the Tour. After a brief delay due to threatening weather conditions, the Tour continued east to Lindbergh Field at Kalamazoo, where they were welcomed by some 10,000 people who had waited for hours in freezing winds. After lunch, they took off for Detroit on the final leg in order of their standing, with Livingston first. Conditions deteriorated so badly on their way into Detroit that right until they sighted Ford Airport at Dearborn, from where they'd taken off so many days ago, the planes were skimming treetops with visibility down to only the length of a city block. Even the presence of President Hoover hadn't stopped many of the crowd leaving as the afternoon grew late and the weather closed in. The field was so muddy when the competitors touched down in steady drizzle and almost in darkness that it took all of the pilots' experience and skill to keep their wheels from bogging in.

Conditions continued to worsen after the Tour was finally on the ground; most of the fliers couldn't take off again for several days. Although he'd been second to land after Art Davis' WACO, John H. Livingston had scored the highest number of points and was thus the winner of the race, the Edsel Ford trophy and $2,500 in his 225hp WACO Straight-Wing biplane powered by the new J-6 Wright Whirlwind engine. Only five of the twenty-nine competing planes didn't made it home, a true commendation for aircraft engineering of the time. Though ranked eighth in points score, Jessie was the best in her class of plane and the first woman to finish the race, qualifying for a $500 prize presented to her at the reception and dinner in a decorated hangar that night.[18] Though the male pilots tended to see the Tour through rose-colored glasses as the years went by, as one of only three women in the race Jessie would later recall that it wasn't all "brothers of the air" during the event; on at least one occasion, a male competitor unsuccessfully attempted to lead her off-course while ostensibly guiding her through unfamiliar territory.[19]

As a result of her achievements in the two major competitions of the previous year, Jessie gained a well-paid job towards the end of 1929 as chief test pilot with the Victor Aircraft Corporation based in Mt. Holly, in New Jersey. Victor had been essentially set up in the last year of World War I by aircraft engineer, designer and pilot Albert S. Heinrich, who in 1910 had flown the first American-designed and -built monoplane, to manufacture his single-seat Heinrich Pursuit fighter plane for which the U.S. government had promised him a contract. But prototypes had to be built and tested and Victor had only built four planes when the war and the demand for the airplane ended, and so the company had remained small. Reports at the time claimed that Jessie was the first female test pilot in the country and that she was planning flights to the West Coast and south to Florida in a race plane built by the company, possibly with design input from her, and she carried out a prototype test flight at the beginning of December. This plane seems to have incorporated Heinrich's 1925 patented design feature of sliding auxiliary wing sections that acted in a similar way to modern wing flaps.[20]

However, by the early 1930s the Depression was starting to bite and her old flying friend Bill Lancaster, who was just another gypsy pilot in the U.S. these days, was finding it very difficult to obtain work. He was living at the Army and Navy Club at Jessie's expense while Jessie had her own apartment in New York at West 56th Street and was still sending a regular monthly allowance of £30 back to Kiki in England. While Bill was just another unemployed male pilot, Jessie fortunately not only had an excellent flight record on her résumé, but she had a gender advantage: she was a qualified woman pilot at a time when they attracted public attention and publicity because they were still rare. In fact, it was just as well she had that advantage; shortly after she obtained the job at Victor, they become another victim of those harsh economic times and closed down. Before she knew it, Jessie was hunting for a job again. To complicate matters further, her mother had arrived in town from Australia at the beginning of the year and moved into Jessie's large apartment, which wasn't cheap, and it wasn't long before the strain of having another mouth to feed began to cut into her tight budget. With that special sixth sense only mothers have, too, Mrs. Beveridge didn't approve of Bill, which made socializing rather awkward.[21] Then Fate played Jessie another hand just in time.

One night at a cocktail party, they met a Dutchman named Charles T. Stork whose company had opened a palatial office as agents for Cirrus engines, with which both Jessie and Bill were well acquainted, as well as Irving parachutes, Stinson Junior monoplanes, Great Lakes Sports Trainers and Savoia Marchetti float planes. Stork needed demonstration pilots who could also double as salespeople and he offered Jessie and Bill $600 a month each, along with respectable commissions on anything they sold.[22] They would consult the list of prospective customers in the office every morning, call them and make appointments and then go sell them an airplane. Because Bill and Jessie would be taking customers up in planes for test flights, Stork's first question when he interviewed them at his office was whether they had commercial licenses in order to carry passengers. To the honest Jessie's horror, before she could admit they didn't have one Bill cut in and reassured Stork that they did. So, they agreed on a start date in about a week, shook hands, and Jessie and Bill walked out into the lobby and stepped into the elevator. As soon as the doors closed, Jessie rounded on Bill and demanded to know why he'd lied. Bill replied that he had a simple plan. During the Ford Tour, he'd met J.R. Booth from the Ottawa Flying Club, who had said to call him if they ever needed anything. Bill was going to wire him right now to tell him that he and Jessie would be on a train that evening to Canada and to ask if, in the meantime, he could make arrangements with the Canadian Air Force to give them the necessary tests for a commercial license. Now, when Bill said they were going to board a train that evening, it was already late afternoon in New York City in the middle of winter. They didn't even have time to go back for a toothbrush. Hailing a taxi, they drove straight to Grand Central Station.

While Jessie was wearing a heavy suit with a wide fur collar, she was concerned her outfit wasn't going to withstand Canadian temperatures, but when the train pulled into the platform the next morning at eight-thirty there was good old J. R. Booth holding a fur coat that went right down to her ankles, as well as one for Bill. They needed the coats; it was so wickedly cold Jessie could hardly breathe. Knowing they had to return to New York as soon as possible, J.R. reassured them he had everything arranged and the test would be that afternoon. In the meantime, he took them out for a very long lunch. Consequently, by the time Jessie walked into the doctor's office for her medical test she was a little under the weather, and when he asked her to read off the letters on the eye chart they were all just a tad fuzzy and then she had to confess the reason for her condition. The doctor just shook his head and decided to carry out the rest of the examination and then they'd get back to the eye test right at the end. Fortunately, by that time Jessie had sobered up enough to see clearly and she passed her medical. Then she took her pilot's test in a small training plane and passed that as well, so in this strange and wonderful way, on 5 February 1930, Jessie became the first woman to obtain a commercial pilot's license in Canada. According to Royal Canadian Flying Club Association figures, only sixteen women in Canada at that time had a private license, so once again Jessie was part of a small, select group of women.[23]

With no time to celebrate this historic moment, however, they hurried to the railroad station, boarded another train, and forty-eight hours after they left New York Jessie and Bill were back there still wearing the same clothes. Their job with Stork lasted some six months, during which Jessie personally sold a Stinson Junior to a New York Oriental goods importer. Otherwise known as the SM-2 Junior, this was a single-engine, strut-braced, high-wing monoplane with an enclosed cabin that could carry two or three people at an average speed of around 113 miles per hour. Jessie had actually been personally passed to fly the Stinson by the company founder, Eddie Stinson, and she knew his sister, Katherine, who in 1912 had been only the fourth woman to gain a pilot's license in the U.S. and who had inspired Eddie to become a stunt pilot before founding the Stinson Aircraft Company.[24] Eddie was very particular about qualifying pilots for his planes and so Jessie had gone to Detroit, where Stinson had personally taken her up so he could watch her fly and then she had to do forced landings in a field. Afterwards, though, he wrote her a letter saying that she flew as well as any man he knew.[25]

Just before they left Stork, Jessie also had a chance to sell a Savoia Marchetti S.56B amphibian, an airplane that had wheels for a ground landing, which could be manually retracted inside floats to convert it into a seaplane. With an enclosed cabin and powered by a 125hp Kinner K-5 radial engine, the S.56B had just gone into production in the U.S.; its earlier model had quickly become the New York City Police Department's favorite airplane with which to catch rum runners. So, when a young man drove up in a gleaming long-nosed Duesenberg to the seaplane mooring dock at Long Island Sound, Jessie knew this was no average customer and there could be a big sale, but there was just one small problem: Jessie hadn't actually been up in one of these planes before. Unobtrusively, she quickly flicked through the instruction manual. They started up and rolled down the slipway out onto the water, but then, to retract the wheels into the pontoons in order to take off, Jessie had to pull up a lever much like a hand brake and then push it over to lock the wheels up. Well, Jessie weighed barely 100 pounds, and even with both feet off the floor she couldn't throw enough weight on the stick to push it over.

With the plane drifting down the Sound, she ultimately had to give up and ask the customer for help, telling him she'd sprained her wrist, and so finally they were able to lock the wheels and lift off the water. The Savoia Marchetti handled beautifully in the air and after a while Jessie handed over the controls to the customer and was pleasantly surprised to discover he was as good a pilot as he was to look at. Coming in to land, she took over the controls again only to discover that the surface of the water was so clear and smooth she was having difficulty judging the distance between her and it. She'd have to throw something in to ripple the water and make it visible, but it was a new plane and there was nothing loose in the cabin to toss out. With a sigh, she took out her powder compact from her handbag and threw it out the window, achieving the desired ripples to judge distance and make a safe landing. The customer was very impressed, either by Jessie or by the Savoia Marchetti, and promised to buy it.[26] Those planes didn't always behave so well. In late May, Jessie and Bill were making a publicity flight over the incoming Grace Line passenger steamer *Santa Elisa* in one when they were caught in a storm seven miles out to sea and forced to make a landing on water, where they then had to sit out the night until they could be towed in the next morning.[27] However, by then Jessie could see that Mrs. Stork had her eyes on Bill more than on Mr. Stork, and she figured it was about the right time for them to make a diplomatic exit if she was to keep him out of trouble.[28]

By now, Jessie would have read the news that she was no longer the only woman to have traveled by air from England to Australia. At 3:57 in the afternoon of 24 May 1930, the young English aviatrix Amy Johnson landed at Darwin in her de Havilland DH 60 Gipsy Moth, *Jason*, having left Croydon in England on 5 May, becoming the first woman to pilot an airplane solo over that route. Having taken up flying as a hobby, she'd only obtained her pilot's license the previous year and had all of about 90 hours' flying time.[29] In honor of her achievement, she was awarded the No.1 civil pilot's license in Australia, which, however, didn't mean she was the first woman in Australia to obtain one, nor was she an Australian woman pilot, of course. Jessie was now no longer unique; although she would always be the first woman by air to Australia, Johnson had just relegated her to history before she'd turned thirty. As the *Adelaide Chronicle* so aptly put it,

> No one will grudge Mrs. Keith Miller the distinction of being the first aviatrix to demonstrate that sex is no disqualification for covering with an aeroplane in a single flight the same tremendous distance. But what has stirred the imagination of the civilised world is that the feat should have been accomplished by a woman able, like Coriolanus, to say, "Alone I did it."[30]

Jessie must have been even more conscious now that she had to do something significant again to reestablish her standing as a record-setting aviatrix in the public mind.

Seven. Flying the Ford Tour

So, she went back to considering more possible record flights and in early June announced that at the end of the month she would attempt to become the first woman to pilot a plane across the Atlantic solo, from Newfoundland to England.[31] Although Amelia Earhart had traveled that route in the air in June 1928, she had been just a passenger in the *Friendship* with pilot Wilmer Stutz and co-pilot Louis Gordon, and as yet no woman had followed Lindbergh's 1927 solo, non-stop transatlantic record. However, for some reason, probably lack of a sponsor or a suitable aircraft, Jessie's record attempt never eventuated. Those factors were certainly an issue shortly after when Jessie heard that fellow pilot Laura Ingalls, whom she'd flown with in the 1929 Derby, was planning to be the first woman to establish a U.S. transcontinental speed record. Perhaps a little older than Jessie, Laura Houghtaling Ingalls came from a wealthy family, and her brother was married to J.P. Morgan's granddaughter. A dancer and musician until she'd caught the flying bug, Laura had promptly quit the stage, taken flying lessons and qualified for her commercial pilot's license in 1929, and then become the first woman to qualify for a transport license in April 1930. Since then she'd set records for flying loops and barrel rolls, but it wasn't the acrobatics that concerned Jessie: it was Ingalls' announcement about her intended transcontinental flight that grabbed Jessie's indignant attention. After all, Jessie had just flown the same New York to California route solo the year before on her way to the Women's Derby and no one had called that a record!

Of course, to do anything at the moment she'd need one particular item she didn't have right now: an airplane. Anyway, at this time Ingalls' flight wasn't a priority because Jessie now had her sights set on a transpacific flight to Japan, so with that agenda she paid a visit to Guy Vaughan, general manager of the newly renamed Curtiss-Wright Corporation in Buffalo, New York, to ask him if they'd be prepared to help her out with an engine. This was the company that had developed the famous Wright Whirlwind and Cyclone engines, and Jessie had known Vaughan, who had started out as a Wright engineer, for some time. If she could persuade Vaughan to sponsor her with an engine, the most expensive part of the airplane, then perhaps she would be able to afford her own airframe to fit out. Vaughan, however, was horrified Jessie was even thinking of such a venture, protesting that she had no idea of the weather conditions or what it would be like to fly alone over water for that length of time. She'd end up drowning herself, he declared. Despite Jessie's pleas, he was firm that he'd only lend her an engine on the condition she flew over land. Thinking quickly on her feet, she recalled Laura Ingalls' announcement. Already familiar with Ingalls' route, Jessie was sure any record she'd set could be quickly broken and, of course, she'd be flying over land all the way! So, with what must have seemed a surprising about-face to Vaughan, Jessie promptly agreed to his conditions and became the temporary owner for six months of a magnificent Wright Whirlwind engine.[32]

Eight

Lost in the Bermuda Triangle

Now that she had an engine, Jessie started a search for a cheap airplane that nobody wanted, eventually hearing about one that was not only unwanted but downright feared—an Alexander Bullet! Built by the Alexander Aircraft Company in Colorado Springs, Colorado, the Bullet had already killed two test pilots and had been consistently refused a Certificate of Airworthiness until five models had been tested. "Just my kind of plane," Jessie thought, and hitched a ride on the first aircraft she could find that was landing at Denver.

The Alexander Company was already famous for their Alexander Eaglerock, a versatile, durable and very popular biplane used for everything from barnstorming to carrying airmail. So popular was it that from 1928 to 1929 Alexander was reputedly the most productive aircraft manufacturer in the world. However, their effort at building an enclosed-cabin, low-wing monoplane, the Bullet, advertised as being able to carry four people and a dog at 120 miles per hour, had been a different story altogether, and only thirteen would ever be completed and registered. Unlike most monoplanes of the period that had wings attached at or near the top of a fuselage that had to be strengthened to take the weight, the more contemporary Al Mooney-designed Bullet's fuselage served mainly as a canopy and engine mount and was thus lighter. Having cantilevered wings below the fuselage also made it possible to eliminate struts and have innovative retractable landing gear, and the result was a distinctively clean, streamlined, aerodynamic design. Even protruding door handles were removed. Having been granted an "X" (experimental) license, the Bullet showed promise. Piloted by Edith Folz in the 1929 Women's National Air Derby and by Errett Williams in the Cleveland cross-country efficiency race that same year, the Bullet performed outstandingly well, as it did in other races, too.

However, before it could receive an Approved Type Certificate from the Air Commerce Bureau, now the Civil Aeronautics Authority, it had to be rig-

Eight. Lost in the Bermuda Triangle

orously tested, and one of the requirements was that it had to demonstrate an ability to recover from a spin after it had been made to spin six turns fully loaded. Usually, if you throttle back the engine, pull the plane up into a nose-high stall and kick over the rudder, it will oblige by falling into a spin, but the Bullet had been built too well and wouldn't spin, staying under control. So, Alexander called in veteran pilot Errett Williams from South Carolina. He took his fully weighted Bullet (X741H) up to 7,000 feet and put it into the usual nose-down spin, but by 3,000 feet it was in a dreaded unrecoverable flat spin, with nose and tail level with the horizon, that would quickly cause a pilot to black out. Williams bailed out at 2,000 feet and suffered more damage than the plane, which was spinning so flat it was only losing a few feet in altitude with each turn and fluttered softly and intact to the ground like a falling leaf. The next test pilot, Bill Sylvester, wasn't so lucky. He didn't make it out during the spin and died on impact. James "Shelley" Charles, the third test pilot, parachuted safely after his Bullet (X747H) went into its flat spin, but Pat Love in number four died, too. After that, the Bureau refused to come out to Colorado anymore and so Garland Peed then took a much-modified Bullet C-7(NX309V) to Washington in February 1930 and put it so successfully through its paces that the Bullet was finally decreed a spin-free airplane and certificated. But by then, the damage had been done to sales.[1]

One of the design alterations was to the direction of the wheel retraction: in early models the wheels retracted inboard, which may have affected the center of gravity, whereas on later models the wheels retracted aft. So new was the concept of having retractable instead of fixed wheels that some Bullets pancaked because pilots, even Williams, forgot the wheels weren't down. The good news, on the other hand, for pilots who did live long enough to learn the Bullet's foibles was that this was a fast, durable and innovative aircraft that was only problematic when deliberately forced to misbehave. Jessie would already have been acquainted with it, having seen it so successfully flown in the 1929 Derby by Edith Foltz, who had won second place in the light aircraft class. When it came to flying long distances, too, piloting from within a closed cabin was more comfortable than an open cockpit. Amelia Earhart, experienced like Jessie in both open and closed aircraft, had pointed out that "the added comfort of a closed plane very definitely lessens fatigue, and fatigue must be considered when one is preparing for a long flight."[2]

Once Jessie had persuaded the Alexander Company that she was not going to be doing any fancy aerobatics, they reluctantly pointed her towards Errett Williams' old C-5 sitting forlorn and somewhat bent in the corner of a hangar. Despite being covered in cobwebs and dust and a little scratched up, it was still structurally sound, and so Jessie and the Alexander mechanics cleaned it up and installed the Wright Whirlwind engine. Then she took the plane up and tested it, not only surviving that experience but also returning to earth happy with the Bullet's performance. Still, she was under no illusions as to the plane's idiosyncrasies as she took off for New York, perhaps the last pilot to take delivery of a Bullet from the factory. As she explained some weeks later, "I am trying to put myself over as a commercial pilot. If I can make a flight in an old ship without any of the usual equipment, it ought to be an easy matter to get some company interested in using me as a regular pilot.... I had showed people I could fly a good plane. I wanted to show them I could fly a rotten one."[3] One of the other appeals of the Bullet for Jessie was its long range, which she proved by not having to touch down again until she reached Columbus, Ohio. There, like that list of forgetful pilots before her, Jessie went to put the plane down on the airfield only to be rewarded by that dreaded horrible, grinding crunch and a slide down the runway. She, too, had forgotten to lower the wheels.

However, to give Jessie some credit, this wasn't an easy process in the Bullet; you didn't just flick a switch and the wheels dropped. After taking off, the pilot had to hand-ratchet each wheel up separately by fitting an iron stick that was clamped beside the seat into a pin in the center of the cabin floor. On the top of the stick there was a split thumb-button: to lower the left wheel, you pressed down the left side of the button and that would ratchet up that wheel and when that was up you pressed down the right side of the button and ratcheted up that wheel. While you were doing all that, you were also trying to keep the ascending plane on-course and level while your revs were mounting as the air resistance lessened. Once the wheels were up, you had to remember to lock the stick back and then take it out; otherwise the metal rod interfered with the compass. Then, when coming in to land, you unclamped the stick with one hand while holding the yoke steady with the other, fitted the stick into the pin, threw it forward and waited to feel the wheels spring down. The immediate impact on the wind resistance was much the same as stepping on the brakes in a car: the airspeed fell off and the plane's descent rate increased. As that happened, you quickly pulled the stick back again to lock your wheels in place and then got both hands back on the yoke before you touched down. So, as well as learning to fly a new and quirky airplane that already had a reputation as a killer, Jessie also had to learn a complicated procedure unique to this aircraft just to land it. Still, the mechanics at Columbus couldn't resist giving her a good-humored ribbing about the belly flop while they were straightening out the dents, and just so she wouldn't forget next time they helpfully painted on the instrument panel in big white letters, "WHEELS."[4]

Returning safely to New York without any further incidents, Jessie tried to get used to living with her mother again. Jessie had had to leave her old apartment at West 56th Street when she went to Colorado and move her mother, who'd been with her since the beginning of the year, into another

Eight. Lost in the Bermuda Triangle 139

apartment. After a while, though, Jessie just had to admit that living with her there just wasn't going to work. "Frankly," Jessie recalled later, "it was getting a little wearing having mother because she disapproved of everything." Including, of course, Bill. Jessie had to have her own life now and so, while her mother remained in the old place, she found another apartment for herself on East 47th Street that had once been the top-floor billiard room of a mansion town house. It was smaller but cozy, with wood paneling and central heating, and, most important, it had room for her piano.[5]

The National Air Races were once again scheduled for late August. Whereas in 1929 the various Derby and Trophy races had finished in Cleveland where the Nationals were being held, this year the venue had been moved to the Curtiss Airport at Chicago. There would be two women's races finishing there: the Women's Class A Pacific Derby starting from Long Beach, California, and the Women's Class B "Dixie" Derby starting from Washington, D.C. Some fourteen women initially registered their interest in the Pacific Derby, including Jessie, who was actually involved in some of the pre-race publicity, but then the rules were changed to restrict the type of plane that could be flown to small light aircraft and to ensure that the competitors were followed by a plane carrying a doctor. Jessie, Amelia Earhart, Elinor Smith and three others objected, pointing out that none of the men's races carried these restrictions and, besides, they were all experienced pilots in heavier multi-seat aircraft. When this had no effect, they all pulled out of the race in protest, leaving only a small group of six women to take off from Long Beach compared to the twenty aircraft involved in the Women's Derby the previous year.[6] Gladys O'Donnell won the Pacific Derby while Phoebe Omlie came first in the Dixie Derby. Third place in that race went to Laura Ingalls.

A few weeks later, at six-thirty in the morning of Sunday, 5 October, Ingalls took off for the West Coast from Roosevelt Field, Long Island, in her open-cockpit Gipsy Moth biplane. Three days and nine stops later, she touched down at the Grand Central Air Terminal in Glendale, California. For the last 300 miles she had groped her way through storms and dense mountain fog. With an elapsed flying time of 30 hours 27 minutes, her westward flight was the first to be recognized as a transcontinental record for women pilots by the National Aeronautic Association, who were no doubt inspired by the recent ruling of the Fédération Aéronautique Internationale to accept records set by women aviators. However, Ingalls was accompanied for the entire flight by a two-man service plane, which just added fuel to Jessie's certainty she could do it faster and totally alone. So, a few days later, on 13 October, Jessie climbed into the cabin of her Alexander Bullet, where she sat on cushions in order to reach instruments and pedals, and took off into the west from Curtiss Field, Valley Stream, at the same time of the morning. Her goal was a flight of around twenty hours with three stops, the first of which would be Indianapolis. "I am anxious to get all this preliminary flying done," she commented before she left, "and then maybe next year I shall make the transatlantic crossing. I shall go alone and establish more firmly women's position in the world of aviation."[7] Jessie reached Indianapolis in five and a half hours, took a fifteen-minute break for refueling and continued on towards Wichita, where she expected to stay overnight. However, when she arrived at Lambert Field in Kansas City just after five that evening, some twenty-one minutes ahead of Ingalls' time, bad light, dense fog and rain were setting in and kept her there overnight and into the next day. Finally, she was able to take off about mid-afternoon and fly on to Wichita, from where she flew the following morning to Albuquerque in New Mexico, and then to Winslow, Arizona, where she stayed overnight. Finally, Jessie came in to land at Los Angeles at 10:58 on the morning of 16 October, with an elapsed time of 25 hours and 44 minutes, breaking Ingalls' record by 4 hours and 43 minutes.[8]

By then, Ingalls was in Kansas City on her way back east. Naturally, Jessie promptly announced that she, too, would make an attempt on the record for the return run to New York on Saturday. Later, Jessie would recall that she'd only been really worried twice during that record flight to California. "After flying eleven hundred miles that first day," she commented to journalists, who likened her to Kingsford Smith because of her businesslike attitude to flying, "I found myself west of Kansas City with no place to set down. I had no landing lights nor instruments and I'd never flown at night." Then she had realized that the rugged mountain terrain beneath her after leaving Albuquerque offered no landing place in case of emergency and even if she bailed out she'd be marooned in the wilderness.[9] When Laura Ingalls landed back at Roosevelt Field, Long Island, on 18 October, she was credited with the first official eastward transcontinental solo record flight by a woman, completed in 25 hours 35 minutes.

Jessie vowed she'd beat Ingalls again on the return trip, and with no further fuss she took off from Los Angeles for the East Coast at 1:10 in the afternoon of Sunday, 19 October. As for all her trips, she did her own navigation, drawing her distances onto one side of her maps and her compass courses on the other. She also took careful note of altitude variations because she'd be flying up to heights of 10,000 feet. Then, landing in poor light at dusk in Winslow, Arizona, she ground-looped, breaking a wheel and damaging the tail skid. "I was thoroughly scared," she remembered.[10] There was no spare wheel available, so Jessie had to wait until one could be flown out from Los Angeles on Monday. She was then able to reach Albuquerque in the middle of Tuesday afternoon, where she had planned to stay the night, but her propeller had been bent in the accident and the plane wasn't flying well. Deciding to play it safe, she waited again while it was sent back to Los Angeles to be straightened and strengthened before being reattached to the plane.[11] Then she took off on Friday morning for Wichita, where she arrived late in the afternoon still well ahead of Ingalls' flight time. With an average speed of over 100 miles per hour, she had increased her lead on Ingalls' time to an amazing eight hours by the time she arrived in Columbus, Ohio, on Saturday afternoon, finally touching down at Curtiss Field, Valley Stream, Long Island, on Sunday afternoon, 26 October, with an official elapsed flying time of 21 hours and 47 minutes, ultimately beating Ingalls' brief record by some four hours to be only the second woman to hold the U.S. transcontinental flight speed record in both directions.[12]

Breaking such an important record so quickly brought Jessie a lot of attention, and she was quickly approached by Harwood Smith, John Liggett III and William Bowman, who were associated with an organization known as Aerial Enterprises Inc. that had been formed to publicize Pittsburgh as an air hub. They suggested that this time Jessie could create a record all of her own: a non-stop flight from Pittsburgh to Havana in Cuba, an estimated distance of some 1,250 miles. It would be a rare chance to conquer unknown territory, and not just for women; no one had yet made this particular journey.

An understandably cheerful Jessie standing on the wing of her red Alexander Bullet after landing at Curtiss Field, Valley Stream, Long Island, 26 October 1930, having become the first woman to establish a U.S. transcontinental unaccompanied solo flight record in both directions (courtesy Carruthers Memorial Aviation Collection, Special Collections, The Claremont Colleges Library, Claremont, California).

"This flight means much, we believe," Smith announced to the *Pittsburgh Press* at the beginning of November, "in that it would serve not only to center international attention on Pittsburgh but to foreshadow the day when such hops will be the accepted mode of travel over long and short distances."[13] To take advantage of the publicity surrounding her departure, a flying show and parade would be organized as support events.

In those days of intrepid flying adventures, though, no record remained unattempted for very long. On 9 November, after Jessie and her friend Bill Lancaster landed in her red Bullet at Pittsburgh, they heard that Frank Hawks had flown to Havana from New York in a Travel Air monoplane in 8 hours 44 minutes. While she would not now be the first pilot to Cuba from the East Coast, Jessie could still be the first woman, but she would have to leave quickly or someone would beat her to it. Met by Harwood Smith and Pittsburgh city officials, she and Bill were officially welcomed. After Jessie spoke at the Women's City Club that afternoon, they all retired to the Schenley Hotel to make plans and wait impatiently for good weather. It was not an easy wait; the weather refused to clear and three days later they were still there. However, the layup did leave Jessie time to talk to journalists and gain some added publicity for her flight, although she was evidently still fielding the same type of questions from journalists as she had back in Australia and so felt it necessary to warn this reporter that she was just as tired of them.

"Now please," she began, "don't ask me the bunk." Then she had to explain the expression. " The bunk is like this: Do you like babies? Do you cook? Do you admire the American housewife? What do you think of aviation's future? Things like that." Never daunted, Mr. Fitzgerald could still think of a trivial question to ask: Given her current projected departure date for Havana was the ominous number 13, was she superstitious about that? She certainly was, like most aviators and sailors alike, but she wasn't about to admit it to the press and the general public and so she adamantly denied it, reminding the intrepid reporter that not only had she been born on Friday the thirteenth, but she'd taken off on her recent record transcontinental flight on 13 October. Then she firmly took the wheel and steered the conversation towards more serious aviation matters. "Women are seriously handicapped," she declared,

> by the fact that transport companies refuse them jobs as pilots on the theory that scared passengers will not ride with them. Women can handle big planes as well as men. Strength plays no part in it. And women are more cautious. But we are handicapped by the fact that we get only the planes that first-rate male pilots wouldn't use. Women's best bet in aviation is promotional work. And that's off now because of the general depression.

Still, for the moment there was another record-breaking flight to think about. If the early-morning weather charts indicated clear conditions south

Jessie in Pittsburgh with her Alexander Bullet prior to departure for Havana, Cuba, in November 1930 (courtesy Carruthers Memorial Aviation Collection, Special Collections, The Claremont Colleges Library, Claremont, California).

of them, she explained, the plan was that she would be called at four in the morning by the Pittsburgh-Butler Airport control tower. Harwood Smith would collect her from the hotel and drive her to the airport, where they would wait for dawn, when, if the good conditions were still holding, she would take off on a route over the mountains south to Jacksonville, Florida. After refueling, she would fly on to Miami to pick up her immigration and Customs clearance documents. From there, with a good tailwind, the flight time to Havana should only be some eleven or twelve hours.[14]

But 13 November came and went, and the weather still did not lift. Jessie fretted, waiting up until after midnight and then waking at around four-thirty in case the airfield called her. Finally, the long-awaited call announcing clear weather came on the morning of the nineteenth, and she took off just before eight o'clock carrying a hundred gallons of fuel in two extra gas tanks that had been fitted, leaving a minimum of space in the cabin and blocking virtually any chance of a quick exit with a parachute in case of emergency. With the fatalistic attitude of many a flier of that period, Jessie was known to declare that she only carried a parachute anyway because it made other people feel better.[15] There was no radio in the Bullet, either, nor night-flying instrumentation nor even a turn-and-bank indicator.

To everyone's surprise, Jessie landed back at Pittsburgh later in the day; her path south had been blocked by an impenetrable wall of fog over a southern spur of the Allegheny Mountains. She returned to waiting for the weather, pacing up and down until the morning of Saturday, 22 November, when the phone rang again at four in the morning. Racing to the airfield, Jessie climbed back into her cockpit while the mechanics packed cushions around her, and took off into the dawn sky, only to quickly realize that this time she'd have a different problem: headwind. The farther south she flew, the worse it became, and soon she was using both hands and both feet to keep the unstable plane level. She couldn't even let go with one hand long enough in those terrible conditions to get out of her heavy coat. Slowed down by the headwinds and with darkness closing in, she was forced to touch down at the Navy airfield in Charleston, South Carolina, forty miles off-course and out of fuel. Only on climbing out of the cockpit did Jessie realize she'd been putting so much pressure on the rudder bar that one side of her shoe was completely worn through. She was also covered in milk: her plane had hit an air pocket as she'd taken the top of a bottle. "I came down all right but the bottle of milk didn't," she observed ruefully. One reporter noted that climbing in and out of the Bullet's cockpit was "a difficult and torturous process" and, perhaps not surprisingly, Jessie seemed to have "a combined hatred and keen interest" in her plane, which she consistently referred to as a wreck. "Someday," Jessie observed wryly, hinting at the plane's reputation, "I'm going to take her up in good flying conditions over a stretch of decent country for landing and gain a lot of altitude. Then I shall take both hands off the controls and my feet off the rudder and see what happens. I think she'll go into a beautiful side-slip, don't you?"[16]

Because she had intended landing at Jacksonville, Jessie hadn't made any provision for accessing fuel from the Navy, who naturally didn't have a policy of refueling civilian aircraft. It wasn't until various residents lobbied Rear Admiral McCully that "the gallant little flier struggling with a task obviously too great for her," as one journalist patronizingly commented, was given the necessary supplies.[17] The next day she left for Miami, from where she flew another 220 miles over water to Havana. When she touched down at the General Machado Airport there at 4:47 pm on 25 November, after twelve hours and eight minutes of flying time from a Pittsburgh 1,350 miles away, a large crowd was waiting. "Awful glad I made it," commented Jessie thankfully as she climbed out of her plane, waving to officials, reporters, photographers

That's not Jeremy Irons welcoming a happy and celebratory Jessie to Havana, 25 November 1930, at the General Machado Airport; it's Mr. J. G. Cates of the Curtiss Company of Cuba, after her first flight by a woman from the East Coast of the U.S. to Havana (courtesy Carruthers Memorial Aviation Collection, Special Collections, The Claremont Colleges Library, Claremont, California).

and a wildly cheering crowd acknowledging her as the very first female pilot to accomplish that particular journey. At an official dinner that night at the British Consulate, she presented to General Gerardo Machado, who had been President of Cuba since 1925, an illuminated address on a scroll from the Mayor of Pittsburgh. Machado was so taken with the idea that a few days later he gave her one to carry back.[18]

Jessie's plan had been to start out on the return journey on Tuesday, but once again the weather closed in and she was marooned in Havana for a week, troubled by premonitions of disaster that caused her to lose sleep. "I don't know why it is," an obviously sleep-deprived and depressed Jesse confessed to a journalist, "but something tells me I'm going down. I've had the feeling ever since I crossed on the way over from Florida and somehow or another I can't shake it off." It didn't help, she mused, that she was flying an "un-airworthy crate" rescued from a junk pile, which "anybody but myself would refuse to fly" and with no blind-flying instruments because she couldn't afford them. Although she'd be returning along the same route over water she'd flown on the way in, Jessie reportedly became increasingly worried about the return trip but insisted she'd keep flying solo. There was an inflatable rubber raft in the plane in case she had to ditch, but she was never confident she'd be able to inflate it in time even if she could get out. "Everybody gives me credit for being brave, and to make a go of it," she said,

> and I never let them think otherwise. But really I am afraid, desperately afraid, when I'm over water or mountains or rough country. I got lost in the Alleghanies not long ago and the fog seemed to hang over me like a death shroud. I was terribly afraid. I feel many times like giving it up because I know it's eventually going to get me. But I can't—people would think me a coward. I guess I've just got to keep on until it does get me. Life at its best is short anyhow, so I guess I have no complaint coming.[19]

The longer the weather remained bad, the more pressure Jessie's backers in Pittsburgh put on her to leave, and so Jessie finally took off from Havana for Miami at the end of the week, despite the warnings of friends there and her misgivings. She hadn't completely filled her fuel tanks because Miami wasn't all that far away, and she was going to refuel there to fly on to Pittsburgh. Once up in the air, though, Jessie flew straight into a shrieking headwind, so she took the hint, turned around, landed, filled the tanks all the way and took off again. A Pan American Airways plane spotted her twenty-five miles out, flying very low. Then she flew on out over the water and disappeared.

When she didn't arrive in Miami later that day, a large group of search planes went out immediately: two took off to search from the Havana end of the route while six went out from Miami. They knew Jessie would have fuel for a maximum of nine hours' flying as a safety measure, although the trip should not have taken more than three. If she had been blown off-course by the strong east winds and had to ditch, she wouldn't be able to let rescuers know where she was because her plane wasn't fitted with radio. By the following day, with no sign of her from search planes or ships in the area, aviation officials gave Jessie up for lost although some people offered to keep searching. Even Laura Ingalls flew to Miami to offer her services in the search, warning authorities that they should keep searching and that Jessie was probably down on an isolated island.[20] Bill Lancaster, who still had faith that the woman now regarded as "unquestionably Australia's most renowned air woman" was still alive, wired the British Ambassador to lobby the U.S. Navy to send ships to search the Caribbean.[21]

She was still alive but not where anyone would have expected. In fact, Jessie had been an early near-casualty of the quirky conditions that exist within that loosely defined area now known as the Bermuda Triangle. Despite leaving Havana at nine in the morning in clear weather, she had soon been confronted by strong headwinds. Setting her compass on the appropriate heading, she flew on over the water for some ninety minutes, by which time she was expecting to sight the Florida Keys any minute, but instead all Jessie could see was a lot of ocean. For another hour she kept doggedly flying, but with no sight of land she began to grow anxious. The wind strength had increased and it was now an ongoing battle to keep the cantankerous Bullet's wings level. Still, she believed in her compass and flew on for an hour longer until almost at her point of no return, beyond which there would not be enough fuel to make it back to Havana. It was a terrible decision to have to make. Although cloud and wind were increasing and the sea growing rougher below her, visibility was still good and Jessie expected to see land at any time, but the sea remained empty of land, of ships, of anything.[22]

In desperation, she climbed to an altitude of 7,000 feet, hoping desperately to see something on the horizon, but there was only angry ocean, storm clouds and a worsening gale. After flying for five hours, her feet were numb from the pressure of keeping the plane on-course and level. She'd passed that point of no return now and was committed to flying on, but by now the compass was swinging wildly. Finally, Jessie was forced to admit to herself that she had no idea of her actual position or even whether she was east or west of her intended course, and her hope began to ebb. "I thought it was all over for me," she wrote later, "and I contemplated diving with the ship head-first into the sea to get the agony over with quickly but decided to fight on a while longer."[23] Then, unexpectedly, as she forced her eyes to look away from the plummeting gas gauge one more time, she noticed that on the horizon the sea was changing color from deep ocean blue to the lighter green of shoaling water. Glancing quickly at her chart, Jessie figured she had to be approaching the tip of Florida. Twenty minutes later, to her utter relief and joy, she flew over land and, still thinking it was Florida, she flew east along a coastline. However, instead of the Keys or settlements, all she could see were swamps and lakes. At last, with her fuel nearly exhausted, she sighted a small coastal village and circled it to find a landing site, but there was none and the tide was coming in on the beach. Using the last of her engine power, she pointed the plane's nose above the waterline and pancaked into the thick bushes and undergrowth just behind the beach, somehow staying upright but tearing the covering on the wings. Dazed by the impact, she climbed out of the plane to find herself in the middle of a crowd of local native people. "Natives swarmed around me and I thought for a while I was on a cannibal island," she recalled.

"Where am I?" she asked.

"Andros Island," someone answered.

"Who does this island belong to?" she called out.

"King George!" the crowd responded with a shout, and Jessie knew she was safe.[24]

She had in fact landed at Kemps Bay on Andros Island, the largest of the twenty-six inhabited islands that make up the Bahamas. This time, the flying gods had smiled on her. Jessie had wandered some 143 miles east from her destination of Miami. If she had missed Andros, she would have just kept flying east into the North Atlantic to become an unsolved aviation mystery of the Triangle when her fuel ran out. Andros Island is technically an archipelago rather than a single landmass, consisting of three major islands: North Andros, Mangrove Cay and South Andros, along with smaller islets and cays covered in impenetrable bush and separated by mangrove estuaries and tidal swamps. As Jessie soon discovered, she'd landed on the more remote South Andros from where it was at least sixteen miles to the nearest telegraph station at Mangrove Cay on an island that had no roads, no horses, not even bicycles.

It was still blowing a gale, so gathering up some screw pickets and rope from inside the plane, Jessie persuaded some of the biggest men to lie along the wings to add weight while she tied the Bullet down. Then, around mid-afternoon, Jessie set off on foot along the coast, carrying her map case inside which were her passport and other valuables, and accompanied by a local boy as a guide. As they struggled through sand and across sharp rocks, Jessie could appreciate the severity of the storm through which she'd flown; palm trees were down everywhere, and sections of beaches and cliff tops had been washed away. Where they couldn't find a way through the debris, her guide would lower Jessie by her hands over the edge of a low section of cliff until she could slide to the beach. He'd follow and then they would wade through the water until they could scramble back up onto the path again. Accustomed to flying rather than walking, her feet soon became painfully blistered, and so, sitting down on a convenient rock, the ever-practical Jessie took a pair of pliers out of her case and cut the heels out of her shoes, which she then tied to her feet with handkerchiefs. Then they continued to struggle on through the night as Jessie became more alarmed, not knowing where her guide might

Eight. Lost in the Bermuda Triangle

be taking a lone woman. Eventually, just after midnight, they came to a cottage by the sea that belonged to her guide's brother-in-law, where an exhausted Jessie fell asleep in a bed shared with his wife and family.

In the bright light of morning, she found that they were at Driggs Hill, on one side of the South Bight between South Andros and Mangrove Cay, with the telegraph station still some miles away on the other side across the water. Although the local fishermen hesitated at first to take her in the prevailing conditions, she managed to bribe them to row her across. Needless to say, it was a very tired, wet and bedraggled Jessie who appeared on the doorstep of a surprised Commissioner Elgin Forsythe and his wife, trying to explain who she was. As Forsythe was the representative of the British government on Andros Island, as well as being a boatbuilder, historian, and head of the Bahamas Sponge Growing Commission, Jessie had at last found the person who could offer her official assistance. However, after giving her a stiff drink, drying her out and tending to her damaged feet, Mr. and Mrs. Forsythe had to reveal that their radio had broken down days ago and hadn't been repaired because the bad weather had prevented the mail boat crossing from Nassau. The passage was still too dangerous, they declared, and so there was nothing Jessie could do but go to bed and rest. Dressed in some of Mrs. Forsythe's night attire, Jessie was only too glad to do just that until she was abruptly woken in the middle of the night by loud banging on the front door and voices, one of which was speaking in an all-too-familiar accent. She was just getting out of bed when Mrs. Forsythe rushed into the room, looking embarrassed. There was an Australian at the door, she explained, who had heard Jessie was there and was insisting on offering a fellow countryman his help, but Mrs. Forsythe had told him it would be improper, as Jessie wasn't dressed. The next minute, to Jessie's surprise, through her door came none other than the tall, lean, ruggedly tanned and bearded figure of the great Australian crawl swimmer Percy Cavill! One of the many swimming children of the memorable Professor Fred Cavill, Percy had in 1897 been the amateur world swimming champion over both the 440-yard and five-mile distances, as well as a young Annette Kellerman's swimming teacher, before he'd gone to work in Europe and then the U.S. as a swimming coach. By the early 1920s, he was working at the Royal Palm Hotel Casino in Miami. When the hotel owner, Thomas Peters, then built the Bimini Bay Rod and Gun Club on North Bimini in the Bahamas, he brought Cavill out there as the guests' swim coach, until the Club was destroyed by the Great Hurricane of 1926. Percy had then moved to Fever Cay near Andros Island, where he quietly faded from public view, to become a beachcomber who the natives claimed had a spiritual affinity with the sea. He would never walk when he could swim and there were stories he even slept while in the water. The famous aviatrix who had vanished had just found the famous swimmer who had disappeared.[25]

Hearing that Jessie was desperate to let people know she was alive and well, Percy promptly volunteered to take her across the fifty miles of water to Nassau in his small fishing boat. Forsythe flatly refused to let her go, claiming it was far too dangerous and he'd be blamed if anything happened to her. Catching Cavill's eye, Jessie suggested to Forsythe that at least they could walk down to the beach and look at the boat. By now, it was just before dawn as the trio traipsed down to the water's edge and duly admired the boat. Distracting Forsythe's attention for a moment, Jessie suddenly jumped into Cavill's boat, and off they sailed, leaving poor Forsythe running up and down the beach pleading with her to return. As the sun came up, Jessie and Percy sailed through the morning in his twenty-foot dory, powered by a one-cylinder engine, through squalls, reefs and high seas until they took shelter in a cove where they lit a fire and had a cup of tea. Setting off again, they eventually arrived in Nassau on Monday afternoon, 1 December.[26] There, Jessie was finally able to dash to the telegraph office and send a short telegram, "I am at Nassau and safe," to John Liggett of Aerial Enterprises, the flight's backers, to let him, Bill and the general public know that the aviatrix given up for lost at sea was still very much alive. To her mother she wired: "I am safe. Notify Friends. Love. Jessie." Her mother was naturally overjoyed. "I am so happy I can hardly speak," she still managed to say in New York, "but I never have for a moment given up hope that my daughter would be found. It just seemed as if a voice kept telling me over and over that she would come out of it alive."[27] While Bill stoically commented that he'd never doubted her flying ability or lost his faith she was still alive, the husband who had been left behind and largely forgotten amid all the fuss, Keith Miller, expressed his "tremendous relief" and also declared that "even when hope seemed definitely to have vanished I still thought somehow that she would pull through. She is safe now—that is all that matters."[28] Bill, attempting to regain some control over a situation that had clearly got away from him, stated that he and a mechanic would leave at dawn the next day with fuel to fly to Nassau, pick up Jessie and return to Andros to refuel the plane. "I probably will not allow Mrs. Keith-Miller to fly her plane back here," Bill declared in his best patronizing manner, "but will bring it myself."[29] Really, as if he'd ever been able to tell Jessie what to do.

Lancaster, Liggett and Karl Voelter, Miami manager for Curtiss-Wright, chartered a Sikorsky seaplane immediately and took off from Miami's Pan American Airport for Nassau early in the morning of 2 December loaded with fuel and oil, as Jessie was confident that the plane was still intact and that at low tide they'd be able to wheel it out onto the beach and take off. Picking her up at Nassau, where she was farewelled by Governor Sir Charles Orr and Colonial Secretary William Dundas, they continued on to Andros Island, where indeed they found the Bullet in exactly the same condition as when Jessie had put it down. As they walked around the plane and inspected

Eight. Lost in the Bermuda Triangle

it, they could see that some minor repairs would have to be carried out on the wings, as Jessie still intended to fly it back to Pittsburgh. However, they weren't carrying anything by way of tools or materials, so they decided to leave it where it was for the moment, securely tied down on this remote island, and they flew back to Nassau only to discover that storms of a different kind had broken out in their absence.

As soon as the news had broken that Jessie was safe at Nassau, some skeptical aeronautical experts began suggesting that perhaps she had not been as lost as she had appeared. They found it hard to understand why Jessie had been blown so far east when weather reports indicated that the prevailing winds had been coming *from* the east; in other words, she should have been blown west over the Gulf of Mexico. There was criticism, too, that she'd set a compass heading directly for Miami without allowance for wind drift in those conditions, nor was anyone impressed that the plane carried little instrumentation and no radio. Defending her fiercely on his return to Miami, Voelter had retorted, "After flying over the island of Andros I can say that anyone who thinks that was done as a publicity stunt is nutty.... Mrs. Keith-Miller feels so bad about the flight and the aftermath that she told me she wouldn't enter the United States again unless she did so in her own ship."[30] To make matters worse, bumbling Bill had managed to let it slip in a news conference before he flew out that he would advise Mrs. Keith Miller to accept a $5,000 offer for an exclusive newspaper story, but "only for the purpose of buying a new plane," and it was quickly apparent that the deal had been struck in short order.[31] Jessie's story had appeared the very next day as an exclusive copyrighted to Universal Service, shutting all the other papers out of access to a major news story unless they paid for the rights to it, and that can only have annoyed many of the press corps. Something else that Jessie kept quiet at the time but would become relevant later was that she had been planning to become an American citizen, having remained in the U.S. until then by regularly renewing her six-month visitor visa. The American Consulate in Havana had actually been on the point of refusing her reentry into the U.S. when she persuaded them that she was about to obtain citizenship on her return.[32] So, perhaps the risk of losing her chance to stay flying in America may also have influenced Jessie's haste to leave Cuba before the Consulate changed their mind, even at the risk of her life.

Jessie felt the public's loss of faith in her keenly. As soon as she could, she defended herself, sending a statement to the mainland press that she was "most unhappy about myself. As an experienced flier I feel ashamed to think that I failed to reach my objective. Those who suggested that I might for publicity purposes do such a thing are despicable. I hope that my reputation as a sportswoman is good enough to refute such a thought."[33] At the same time, she reminded people that Bill did not make decisions for her; her choices were her own, for which she took responsibility. "Captain Lancaster did not try to sell my story to the newspapers," she declared. "He was requested by all the news services to state how much the sole rights could be purchased for and he merely advised me in the matter and referred everyone to me. I sold my story as I did not see any reason why I should not recover a small amount of my expenses that I was put to through my flight."[34] Her plane was in good shape, she pointed out, and she would resume the flight to Pittsburgh as soon as possible. By the time she finally arrived back in Miami with Bill and Liggett on 5 December, still dressed in the same clothes and cut-out shoes in which she had walked the beaches of South Andros, Jessie felt justified in claiming she'd been badly treated by the media and that she was reluctant to say anything now lest it be misinterpreted. "I am broken because of what my country thinks," a worn and dejected Jessie announced, using a map to point out where she had gone down. "If you could have seen the situation, there would never have been any doubt."[35] She denied that she'd ever said anything about having a premonition of death or that she had at any time "minimized the worthiness of her plans."[36]

In the early hours of the next morning, Jessie and Bill flew back to Nassau in a seaplane loaded with fabric and "dope" with which to repair the Bullet, picked up oil and petrol and took off for Kemps Bay on South Andros. There they unloaded the barrels with the help of local islanders and spent hours lying underneath the wings restitching the fabric and sealing it. Then they refueled the plane and pushed it onto the beach, from where Jessie at last lifted into the air on her delayed return flight to Miami, finally landing nine days after her original intended date. Resting briefly, she took off on 14 December from Jacksonville, Florida, for Washington, D.C., and Pittsburgh to resume her interrupted flight, watched by Bert Hinkler, his wife and the cartoonist John Held, Jr., who had flown down from New York while Hinkler was teaching Held to fly. But then whatever luck the notorious Bullet had left ran out. As Jessie took off from the airfield at Jacksonville, the fuel pump gave out forty feet in the air. The motor promptly quit and the Bullet plowed into the ground, destroying the wings, undercarriage and any chance Jessie had of completing the round-trip flight.[37] Although the engine was salvageable and she was able to return it, Jessie had to reluctantly abandon the wreck of the airplane that had survived so much and taken her so far. That plane's reputation stayed with it, though. When two aviators later rebuilt Jessie's Bullet and took it up again, it once more fell out of the sky, killing them both.[38]

Bill Lancaster and John Liggett had been returning north by train, but as soon as they heard the news of Jessie's crash they turned around and came back for her. Devastated, Jessie admitted to reporters that she had no idea of her future now: "I have no ship. I hope I get expenses for the flight and what is due on my contract for flying—but I don't know. I'm going to New York tonight to demand an apology for some statements issued against me. I was promised $1,000 and expenses to make the trip but I haven't received it. I don't want to seem mercenary, as I've been falsely painted before, and yet what can I do? I've nothing left."[39]

Her intuition did not lead her astray; events proved she was justified to have some trepidation about being paid. Despite John Liggett's apparent empathy for her plight, Aerial Enterprises subsequently refused to hand over any money because Jessie hadn't completed the round-trip. Furious and in pain from internal injuries sustained when she crashed, Jessie journeyed to Washington with Lancaster, probably to attend a two-day conference there organized by Secretary of Commerce Robert Lamont to discuss uniform air laws and regulations across the U.S., attended by delegates from every state. She and Bill then went to Pittsburgh to tackle John Liggett III and Aerial Enterprises Inc. face-to-face to negotiate a final financial settlement for the Havana flight, with which she was successful. From there, Jessie was returning home to New York for a long-awaited medical examination by her own doctor. When she reached there, she declared, revealing the insinuations about the Havana flight still rankled, "I am going to demand an apology from newspapers for the treatment accorded me after my forced landing at Andros."[40] Her mood may not have been lightened by the news that fellow aviatrix Ruth Nichols had broken Jessie's transcontinental flying record on 10 December, when she landed her Lockheed Vega *New Cincinnati* in New York in a record 13 hours and 21 minutes from Los Angeles. Nichols, who had also been the first to fly non-stop from New York to Miami, was another charter member of the Ninety-Nines. On 19 December, Jessie was admitted to the Murray Hill Sanatorium in New York for treatment for kidney and back injuries she'd sustained in the accident. There she remained in recovery for some months, watched over by her mother.

Lying there recuperating, Jessie might have been cheered up by a letter published back in Australia in the *Freeman's Journal*, if only she'd been able to read it. "My favorite heroine is Mrs. Keith Miller, the Australian aviatrix," wrote thirteen year-old Jean Harper, from an outback New South Wales homestead, Etiwanda, near the small town of Nymagee:

> Mrs. Miller is my favorite heroine because she is so brave and courageous, and dauntless; her spirits do not sink if she is not successful at her first try. I think Mrs. Miller's character is very strong-minded and determined and very rarely excited. And I think her nature is very high spirited and happy-go-lucky and takes the world as it comes. She is always smiling whenever one sees her photo. She smiles also with her eyes.... She can handle her aeroplane very nicely and expertly. Mrs. Miller has flown to many places, and always she has come through smiling brightly, and I think she must have a wonderful nerve to get through the strain of flying from daylight until dark.... Yes indeed, an airwoman is one of the bravest women.[41]

Nine

Hard Times Down South

When Jessie eventually emerged from the Sanatorium, she revisited the West Indies and Cuba for a short while before returning to New York. The Depression had begun to make a serious impact on the American economy and the availability of work, and jobs for aviators were as hard to find as any other job out there. Hoping it might make finding work easier, Jessie began the paperwork in June 1931 that was her first step towards American citizenship. "America gave me an opportunity in aviation that I was unable to get in England," she commented. "I want to become an American because I really want to get somewhere before I die." However, with her usual entrepreneurial spirit, Jessie tried to set up some flying ventures. She'd desperately wanted to be the first woman to fly solo across the Atlantic, but, in April of 1931, Ruth Nichols announced she was about to fly from Newfoundland to Paris the next month. Nevertheless, Jessie announced she had plans for her own transatlantic flight in a plane being built for her, accompanied by a navigator. She wished Ruth well but didn't think she was wise to attempt the flight solo.[1] Their plans didn't eventuate for either pilot, though, and it would be left to their friend

A studio portrait of Jessie taken in New York, c. 1931 (author's collection).

Nine. Hard Times Down South

Amelia Earhart to be the first woman to make the solo transatlantic flight on 20 May 1932.

Perhaps to help her plans or perhaps because he'd just grown tired of waiting for a wife who never returned, Jessie's long-suffering Australian husband, Keith Miller, still living in Melbourne, finally lodged an application there on 25 June for divorce on grounds of desertion, exactly one day after the statutory three years then required for a desertion petition. She had indeed deserted him for the last three years, Keith Miller petitioned the court, without just cause or excuse. Jessie took it on the chin. "He's done me a favor and that's fine of him," she told reporters at Newark, New Jersey, where she was said to be preparing for an air tour of the New England states. "He had home and fireside ideas. I had others. I chose flying and demanded the right to live my own life." Asked if he intended to step into Keith's shoes at the altar, Bill Lancaster merely replied to journalists, who either didn't know he was still married to Kiki or were just ignoring it, that they had expected this and he couldn't give them an answer right now.[2] Bill was apparently quite unsettled by her sudden change of status. "He was furious that I had got mine," Jessie later recalled about the divorce. "He was frightened then that I was free and he wasn't."[3] The divorce was finally granted in Melbourne in November, but it seemed to be a reasonably amicable one, perhaps because by now they had both accepted their marriage was never going to work out.

However, they hadn't reached the end of this relationship for want of effort by Keith Miller.[4] The following year a newspaper in Jessie's home state, Western Australia, claimed to have gained access to some of Miller's letters to his wife, perhaps from divorce court records, and they revealed that he'd tried repeatedly to persuade her to come home, despite his assertion that in 1927 his wife "absolutely declined to be a wife except in name," which meant in the more politely obtuse phrasing of the time that Miller was claiming Jessie had refused to continue sexual relations with him. Understandably, the woman he met in Sydney after her flight from England was a very different woman from the one who'd left Melbourne the previous year, changed by her experiences and by the development of her relationship with Lancaster. According to this journalist, Jessie had told her husband that she "must go on and on. Nothing on earth could persuade her to settle down again to a life of wifely responsibilities." After the crash of the *Red Rose* in Cooma and Jessie and Bill's return to Sydney, where they'd met Harry Lyons, Jessie had evidently wired her husband to meet her there, where she'd shown him her passport already stamped with a U.S. visa and informed him she was leaving. The only address she gave him was that of a San Francisco bank, from which his letters were subsequently returned, unopened. Her mother, Mrs. Beveridge, had tried to intercede on his behalf when she met up with Jessie in New York, but to no avail. Miller then wrote again, begging her to come back.

He had a car now that she could drive and he'd support her flying. He'd send her money for her trip home. "I think this is a reasonable proposition," he reasoned, "and I appeal to your sportsmanship to accept it. You are doing well for yourself. Isn't it up to you to give me some consideration? The present state of affairs is becoming intolerable to me. This uncertainty is causing me a lot of distress. Do please reply." He eventually got his reply, but not the one he wanted. "I am afraid that my career in aviation means too much to me to give it up," Jessie wrote. "I am sorry to have to say that, but I feel I cannot live with you as your wife again, and I hope you will forgive me. I hope you find happiness and wish you the best of luck."[5] With Jessie carried on the wings of ambition, it was herself for whom she should have been wishing happiness.

As her dreams of actually flying those long distances faded, Jessie turned to making money from talking about it, setting out once again on the summer Chautauqua circuit. The Chautauqua, named after Lake Chautauqua in New York State, which was the site of the first assembly, was an adult education movement that had become popular in America in the late nineteenth century, similar to the earlier Lyceum movement. In the days before radio and television and rapid transport, Chautauqua assemblies brought entertainment and culture to small-town communities, with entertainers, teachers, musicians, preachers, politicians, reformers, and travelers to far distant lands recounting their adventures. Having by now gathered a lot of experience on public lecture circuits in the U.S. and back in Australia, Jessie enjoyed entertaining audiences with tales of her daring flying experiences.

Towards the end of July, she was on a Chautauqua tour through upper New York State with the Redpath Chautauqua company when she became an entirely different kind of heroine in Fort Plain, a small town on the Mohawk River between Utica and Schenectady. In the early-morning hours of 30 July, Jessie was awakened by smoke drifting under the door of her second-floor room in the four-story, fifty-room Grant Hotel, in which she was staying. Discovering that the storeroom on her floor was on fire, she ran along the corridor in her nightgown, banging on doors, rousing guests and the night clerk, who turned on the fire alarm and then dashed out and rang the village alarm bell. Due to Jessie's efforts, everyone in the hotel escaped unhurt, although the fire spread so quickly some had to climb out the windows to ladders or escape over the roof. For a while the entire business district was threatened by the blaze until brigades summoned from surrounding townships arrived. Some fifty firemen fought the fire for over three hours, but although they could save neighboring buildings, the Grant Hotel was completely gutted, with the loss estimated at some $100,000. When it was all over, Jessie was hailed as the savior of many lives that night.[6]

That, however, was the high point of her year. Her mother returned to

Nine. Hard Times Down South

Australia, where she revealed that Jessie still had plans for long-distance flights and that, despite her occupation, Mrs. Beveridge had never actually been up in an airplane in her life.[7] Although Jessie planned to compete again in the Cleveland National Air Races at the end of August, she was probably not well enough to do it. Bill had been working as personal assistant to Jack Maddux, the president of Trans-Continental & Western Airlines. Originally a car dealership owner, he had founded Maddux Airlines in 1927, which had become so successful that within a year they were flying the Los Angeles to New York route, and they had eventually merged with Transcontinental Air Transport in November 1929, with Maddux as head of West Coast operations. The following year, TAT merged with Western Express to form the company that eventually became TWA. Maddux liked Bill a lot and, knowing he was out of work, probably gave him the job as a favor, because even Jessie had to admit Bill was totally unsuited for it. Nevertheless, the two men hatched a scheme for a promotional transcontinental flight by Jessie with the actress Mary Adams and her dog as passengers, and Maddux flew Jessie out to Los Angeles.

But Jessie wasn't well at all and in those days when medical diagnosis could often be approximate, to say the least (after all, look what happened to Jean Harlow), it wasn't until her appendix actually ruptured that Jessie was rushed into the hospital at the end of September. There she endured a traumatic operation from which she took a long time to recover while staying in Adams' little Spanish villa. Any plans she had to make another attempt to cross the Atlantic were now shelved indefinitely. Bill hadn't done much better, for that matter. As Jessie had accurately observed, he wasn't suited to flying a desk, and eventually Maddux had to let him go, nevertheless presenting him with a large black Lincoln car as a bonus to soften the blow.[8] So there they were, both now unemployed and stuck in Los Angeles with 2,000 other jobless pilots in the middle of winter and the Depression, "wondering if we would ever feel the 'stick' in our hands again," Jessie remembered later. "I spent days discussing impossible plans with Bill until my head ached with the futility of it all."[9] But at least they had a car.

However, Bill had reunited in Los Angeles with fellow aviator Gentry Shelton, whom he knew from the Ford Tour days. Shelton came from St. Louis, where in 1929 he and airmail pilot Joe Hammer had managed to remain aloft for five continuous days above Lambert–St. Louis Airport and where he'd had a flying charter business that had collapsed. One night he suggested that they should all drive across country to Florida for the Miami All–American Air Races in January 1932, where they might be able to compete or at least have a good chance of networking with colleagues and perhaps finding work. Shelton owned a plane, a Lockheed currently being hired out in New York, and they figured there might be a chance to develop a charter business flying around the Gulf of Mexico. If they pooled their assets, they had a plane in New York, a large Lincoln car, a small amount of money, and three pilots. They had nothing to lose and everything to gain and so, taking turns driving, Bill, Jessie and Shelton drove across country to Miami. The closer to the city they drove, the more it seemed like L. Frank Baum's Emerald City of Oz in Jessie's mind. Here they would find the answer to what was missing in their lives, she thought. Here they would be able to build up a small aviation business that would carry them through these hard economic times until she could go back to work breaking aviation records and being a test pilot. After all, as far as Jessie knew at that time she was the only female test pilot in America. Jessie could already see the colorful posters advertising their joy flights. Carried away with their enthusiasm, they stopped the car in the middle of the hot Arizona desert and got out to solemnly shake hands in partnership, swearing to sink or swim together in Miami, where "we would ride out the Depression until the flush days came again with the blue skies above us and glorious days."[10]

When they arrived they met up with Jack French, who was to be their business manager. While at the Air Races one day, they were introduced to Mark Tancrel and Jack Russell, who were starting up a company called Latin-American Airways and who in the course of conversation suggested they might employ Jessie and Bill for charter work if they could find an airplane. Tancrel and Russell claimed they were about to travel west to set up some routes between the U.S. and Mexico. Bill figured it sounded like a good idea, even if he wasn't entirely convinced about Tancrel's credentials, and so he said he'd be happy to give the matter some serious consideration, that he knew where there was an airplane they could use and so perhaps they could put something together when he returned from some charter work back east. All in all, Miami looked like a promising place to be, and so Jessie set about looking for a place to rent while Bill was away. She eventually found a large, light and airy two-story Spanish Mission–style house with a red terra-cotta tiled roof and its own garage in an area near Miami known as Coral Gables, at 2321 SW 21st Terrace, set within nearly an acre of land. There was a small orchard of grapefruit and lemon trees and a beautiful garden of bright flowers and scarlet hibiscus. "The windows were framed in purple bougainvillea and the trailing jasmine vine," she wrote later. "At night time, its scent would drift through the open windows and fill the upstairs sleeping porch ... with its overpowering sweetness."[11] It was the first house on its own land in which she had lived since leaving Australia all those years ago, and it would be the first and only residence in which she and Bill lived together, although they didn't share a bedroom. While they still cared for each other deeply as friends, they were not in a steady romantic relationship by now and had not been for some time.

The ground floor of their house contained a central lounge room with a fireplace, from where an archway led into a small dining room and a kitchen at the back, along with a large bedroom and en suite bathroom. As soon as they moved in, Jessie rented that lower bedroom to French. Upstairs was another main bedroom with adjoining bathroom that Jessie immediately claimed as hers and which became her retreat. "I was able to lock myself in away from everything whenever I wanted to," she recalled. "I would read or write or just go there to get away from all the flying talk for a little while."[12] Beyond her bedroom was a large screened porch or sleep-out that accommodated at least five single beds, some closets and bedside tables, and this became Bill's bedroom when he was in town. Jessie thought the area would also be ideal to accommodate various flying buddies or it could even be converted into a self-contained apartment if the need arose. Indeed, no sooner had they settled in than those flying buddies came to call and to stay on. Of course, like Jessie and Bill, they were all unemployed and didn't or couldn't help with rent and expenses. Soon, the house seemed constantly full of unemployed airplane personnel and all Jessie could see was the little money she had saved being rapidly depleted. Quite understandably, she began to feel rather resentful that Bill apparently expected her to keep the house going by herself while feeding all and sundry.

Not long after they arrived, Jessie was driving a friend home one night after a few drinks when she was involved in a minor accident in which she rear-ended another car. For one reason or another, the police were called in, which was potential trouble for Jessie. An arrest for driving under the influence could impact on her pilot's license, so although Bill had actually been sitting in the backseat he claimed that he'd been driving. The police promptly arrested Bill, although he was only in jail for an hour, fined him $50 and gave him a suspended sentence. Given that this was still Prohibition and that Bill and Jessie, like many others, had their own still in their basement, it's probably just as well the police took only a cursory interest in their case, but again Bill didn't know when to leave well enough alone and bragged that there hadn't been enough evidence for a conviction. In words that would later come back to haunt him, he wrote in his journal after the incident that "American justice [was] all wet."[13]

Meanwhile, Shelton had left for New York to pick up the plane that was to be the main asset for their business, but as the days went by there was no sign of him returning. They were fast running out of money and Jessie became understandably worried and depressed about their future, arguing with Bill about their lack of money and inability to pay the rent. "Chubbie raises hell about lack of cash. Do not blame her," Bill succinctly observed in his journal on 13 January.[14] He now had to resort to a bit of literal hunting and gathering to put food on the table, keeping up a supply of chickens, rabbits and ducks pilfered from neighboring backyards. It was clear that something had to be done. In desperation, Bill began his own paperwork to apply for American citizenship, but that didn't solve the problem of the missing Gentry Shelton and their plane. Finally, Shelton called them to say he'd had to buy another plane, a used Curtiss Robin monoplane, as the Lockheed had been wrecked in an accident. However, he'd injured his hand and couldn't fly it down. Bill would have to come get it, and so Bill boarded a bus for New York.

Jessie didn't appreciate being left behind in Miami with neither man nor money, as Bill found out when he telephoned her on 26 January from New York. "She disappoints me greatly by her failure to be her sweet self for a few moments. Did she want cheering up as things are still black. It's a dog's life," he whined to his diary.[15] New York in winter is not a city in which to be either broke or lovelorn, and as he resorted to borrowing money an anxious Bill drifted closer to his emotional edge. "Miss Chubbie terribly," he agonized at the beginning of February. "If anything has happened to her I shall suffer as I have never suffered in my life.... I love her more than my very life. I think she needs me. If I did not think this, I would give my life to make her happy." It simply didn't seem to occur to Bill that Jessie didn't have the same feelings for him any longer. "I was cooling off," she recalled later. "I thought there was no future for us."[16] The reality was that Jessie didn't need him any longer. She was now a famous pilot in her own right, which Bill was not, and she'd been advised more than once that Bill was actually holding her back. If she wanted to progress her professional career, her advisors had counseled, she'd have to do it without Bill.[17] However, the Depression had intervened and now Jessie was being a loyal friend, an attitude that Bill evidently saw in a different light. He never openly admitted that Jessie's feelings for him may have changed, at this or any other time. He seemed to think that if he didn't say it, then it didn't happen.

Eventually, Bill located their Robin plane and the missing Shelton, who, it turned out, had slashed his hand so badly he was unable to handle flight controls. So, leaving him in New York to recuperate and follow later, Bill flew the Robin home to Miami instead. Jessie, still unhappy, was not at the airfield to meet him, a sign of things to come. While Bill had been in New York, Jessie had unsuspectingly arrived at a fork in the path of her life to find someone else waiting for her. A publisher had contacted her offering to buy her autobiography, but Jessie was daunted by the size of such a project. Although she'd written published accounts about her life already, they'd been articles for magazines or newspapers; she'd never worked on a book, and so she felt she needed help. Then Fate stepped in. Having been something of a local "famous flier" celebrity in Miami since her Cuban flight adventures, Jessie was often invited to speak at functions, and at one of these she met the slim, strikingly attractive widow Mrs. Ida Clyde Clarke, a well-known writer and politically connected supporter of women's rights who was currently a journalism lecturer at the nearby University of Miami. Little did Jessie know, but she had encountered a woman with the reputation of being a force of nature equivalent to any storm through which Jessie might have flown a plane.

Born into a Meridian, Mississippi, family with newspaper ink in its veins, the university-educated Ida Clyde Gallagher had surprised no one by joining the editorial staff of the *Nashville Tennessean* and then the *Nashville Banner*, where she married one of the editors, Thomas Clarke. They had two boys, Beverley and Charles Haden, but sadly, Thomas then passed away after a short illness in 1911. That same year, Ida co-founded the Nashville Equal Suffrage League along with Anne Dallas Dudley, who was instrumental in persuading Tennessee in 1920 to become the thirty-sixth and deciding state to ratify the Nineteenth Amendment and thus give American women national suffrage. Ida then moved to New York and on to national and international women's movements, becoming first president of the Business Women's Equal Suffrage League, a member of the National American Women's Suffrage Association and the press representative of the National Council of Women. Ida didn't only campaign for women's rights; she also wrote about them as founding editor of *The Independent Woman* (later *National Business Woman*), contributing editor of the *Pictorial Review* and, by now known as the "feminine Will Rogers," author of six books, two of which, *American Women and the Great War* (1918) and *Uncle Sam Needs a Wife* (1925), are still regarded as early classics of feminist literature encouraging women to achieve fulfilled lives in their own right. In 1927, Clarke had named three prominent women, Ruth McCormick, Alice Longworth and Anne Morgan, as possible candidates for the presidency of the U.S. If not the first American female author to name other women as candidates for the presidency, she was at least one of the earliest.[18]

A few days after that first meeting, Mrs. Clarke invited Jessie to her Miami Beach home for afternoon tea, where she queried her about future plans. They discussed the dire state of current employment for pilots and then Jessie mentioned her book offer and her concerns, although she was sure there was a story to tell. After all, her friend Amelia Earhart had successfully published two books by this time and Jessie felt she certainly had enough of her own flying stories to work into a book. As it happened, Mrs. Clarke had an answer to Jessie's problem: her son Haden, who she claimed was a published writer. Jessie protested she wouldn't be able to pay him, but eventually they worked out a plan whereby he could board with Jessie while they worked together. That Jessie allowed herself to be talked into this arrangement also says something about her loss of self-confidence by this time: Jessie would actually have had a far larger body of work published than Haden Clarke, yet she now felt the need of a stranger to help her write about her own life, something she'd been doing for years.

When Haden and Jessie met face-to-face for the first time a few days later, Haden turned out to be clever, quick-witted, well-educated and very charming, and they soon settled on a work arrangement whereby Jessie would write down the rough draft of her experiences as she remembered them and then Haden would type up the manuscript from her handwritten pages, discussing any changes together as they went along. At first Jessie didn't really like him, but by the time Bill flew back to Miami in the Robin at the end of the first week in February, Jessie and Haden were getting along well. Once he'd heard Jessie's plans, Bill was just as enthusiastic about the book idea, and he took her out to a movie to celebrate: Lionel Barrymore in his latest film, *The Man I Killed*. One can't help wondering if Jessie or Bill ever remembered that film later. So it was that on 12 February this tall, dark young man with piercing blue eyes and a cynical sense of humor came to live in the upper-floor screened porch of the house in Coral Gables. "Little did I realize that sunny day when I first met him," Jessie would reflect later, "that he was to be the cause of so much grief and suffering in my life."[19]

Haden Clarke, Jessie's fiancé (author's collection).

At first Haden seemed as enthusiastic about the project as Jessie, but there was something about him she couldn't quite put her finger on that set off warning bells deep within her female instinct. Without doubt, Haden was an affable guy, and as time went on he and Bill began to spend more time together. Consequently, work on the memoirs slowed to a crawl. Jessie once again felt that she was being left to do everything about the house, including paying for everything, but there was no money with which to do it. To make matters worse, now neither of the people who were supposed to be helping earn that money were contributing. Understandably, Jessie started to lose her temper with Bill, as she had to constantly remind him that they were supporting Haden because he was supposed to be working, which he wasn't doing if Bill took him out for the day! More than ever now, Jessie was tired of Bill's constant harebrained schemes that never amounted to anything and the worthless flying buddies who were always hanging about the place, getting drunk and making a mess and never seeming to leave. "I was pretty acid about the whole thing," she later recalled, an attitude that either went straight over Bill's head as usual or about which he was just in denial while blissfully observing in his journal: "Chubbie running the house and doing most of the work, God bless her."[20]

At other times, Bill would escape from the house with his fellow fliers, leaving Haden to sit around drinking homemade moonshine with Jessie. When the three of them were together, they drank and talked, drank and ate, drank and fished. When he had some money, Bill took them up in the Robin to see Miami from the air. When there wasn't any money, they would happily steal the occasional chicken from a neighbor's coop to cook over a backyard campfire, because by now the power company had cut off their electricity. Bill always tended to take a fatherly attitude towards Haden, who alongside the older pilot always seemed as young as he really was. Jessie didn't know he was actually five years younger than her because Mrs. Clarke had implied he was older.[21] It was the same attitude that Bill always adopted towards Jessie with that paternal, rather condescending way of referring to her as "child: or "kid" or by her childhood nickname of Chubbie. Now faced with the situation that Jessie was free and he wasn't, Bill was both furious and frightened. He simply refused to believe the very idea that Jessie might have a mind of her own, apparently convincing himself that if only he could obtain a divorce from Kiki everything would turn out right.[22] But Kiki knew Bill far too well, and she made it absolutely clear that no such thing would happen unless she was holding £10,000 right in her hand. Mrs. Lancaster knew her man and so she also knew that Bill was no longer the airman who could earn that amount of money himself. Perhaps if it all failed for him in Miami, he might return to her and the children. Bill's only source of money in these hard times was Jessie, and she was tired of his vacillation over his domestic situation.[23]

So, Bill Lancaster was a man caught between those well-known landmarks of desperation: the rock and the hard place. Jessie might have turned down those solo offers before, but now she was free to take them. She'd even received one or two marriage proposals already that "were shooed off by Bill—hands-off business," she recalled. "I could see this going on indefinitely."[24] Bill's possessiveness was part of the problem or, rather, his inability to take possession. Now there was another man about the house, Bill grew even more obsessive and possessive about Jessie, refusing to let her out of his sight and escorting her everywhere. Bill wanted her for himself, but he wouldn't divorce his wife and formalize his relationship with Jessie, a lack of will that was ultimately significant for Jessie and much more significant in the society of the 1930s than it is now. On the other hand, Bill was always turning up everywhere she went, never allowing Jessie to have her own life. "They used to call him my 'old man of the sea,' " she remembered, referring to Coleridge's Ancient Mariner; "I couldn't get rid of him." Deep inside, she longed for the very things Bill couldn't give her: security, marriage and a chance to settle down. "He was around," she added, "and I put up with him but I wanted to do something with my life."[25]

While Bill had been in New York, Jessie had been approached again by Tancrel and Russell, who were in the process of setting up their Latin American Airways company, and when he returned they all met up. They needed to investigate their route in more detail before flying it commercially, they said, insisting that Bill join them on the trip to Mexico, along with Gentry Shelton, who had now recovered. They even gave Bill a small advance on his salary, which he immediately handed to Jessie. It seemed all too good to refuse and so Bill said his good-byes, had a heart-to-heart with Haden about keeping Jessie sober and looking after her, and then flew west with the men.

By early March, they were in St. Petersburg, Florida, from where Bill wrote to Jessie on 6 March, enclosing a small amount of money with the letter. As if she wouldn't know, he reminded her how much gas she could buy with it, ending the letter by commenting: "Sweetheart, what a couple of plums you and Haden turned out to be." That night they reached Beaumont, Texas, from where Bill wrote again to Jessie, "My darling little sweetheart, give my very best to Haden. Tell him I'll write him a note from Nogales. Don't leave my letters lying around. They are for you only. Good night, Love, Bill." Three days later, though, they were still in Beaumont and Jessie hadn't replied. Bill wrote her that "as soon as I can beg, borrow or steal, I'll send some cash to you." But when he finally did hear from Jessie, she was not happy. "Bills are streaming in from everywhere," she complained. "It is hard to write an airy, optimistic book not knowing where the next meal is coming from. I cannot get the laundry. I have pawned my watch to get food. I had to eat the hamburger I bought for the dog." And then she added a revealing line. "A very passionate love letter arrived for you from a girl named Lola. You should not have told me to open your mail." Either Bill had flying groupies or perhaps he was backing another horse in the race after all.[26]

Despite her difficulties back in Miami, though, Jessie and Haden had made all kinds of good resolutions to work on the book. They would sit down and she would tell him her experiences; he'd take notes and then go up to his room and shut the door to work on his typewriter. But after a while, Jessie would become aware that the clacking of keys had fallen off into silence. She didn't know, she would claim later, that Haden needed a toke to type; she'd had no exposure to drugs and so had no idea of their effect on people. Haden didn't smoke marijuana when he was actually with her, although she'd heard him talking with friends about it and they'd unsuccessfully attempted to persuade her to try it. "You have got to remember this, that I was absolutely naïve," she protested years later. "I was an absolute fool when I was young. I led a terribly sheltered life. I had been brought up so strictly."[27] After a while, unable to bear the silence any longer, she would climb the stairs to find him just sitting, smoking and gazing out of the window. She'd ask how much he'd done, to which Haden would just reply that he couldn't write on demand; he had to wait for inspiration. Jessie would sigh with frustration, but she could never be angry with him for long. He was so charming with that dark, wavy hair and big blue eyes. And so, although her first impressions had been right in a way, a vulnerable, bored and lonely Jessie found herself falling for this smooth-talking man who would make such a fuss over her on those warm nights with a big moon and stars above them and the scent of jasmine in the air.[28]

Unlike Bill, Haden treated her as an equal and as a woman who was worth romance and love. He was thoughtful and interested in her opinion, in her clothes, in her hairstyle. No one had ever taken this kind of interest in her before and Jessie was really touched by it. While Bill talked only of flying, Haden was well educated and widely read; one of his friends was writing a book about Rasputin and he'd invite people around to read poetry together. With no money and little food, he and Jessie lived as if marooned on a desert island, surviving adversity together. Their shared misery drew them closer. They'd dream together about spending the money they would make from the book traveling on a yacht to faraway places. Then one night as they were walking home after a dinner party with friends, with that big, bright moon reflecting on the water, Haden kissed her and Jessie decided she must be in love with him and, before she knew it, Haden had asked her to marry him and she'd said yes.[29] Much later, she would attempt to backpedal and justify her "infatuation" as happening because they were "living under a terrific strain," to escape responsibility by claiming Haden had been spiking her drinks, but that was a long time after she'd penitently thrown herself on her sword to save Bill. Like most people, Jessie probably didn't want to remember that she'd once been vulnerable and had completely fallen under the spell of someone else.[30]

Unfortunately, as it so often does, reality hit them the next day, hard and with more than a twinge of guilt. How on earth were they going to tell Bill? After all, he'd been a good friend to them both and, of course, he and Jessie had been through so much together. Despite his faults, she'd always regarded Bill as a brave and talented pilot who never panicked when things went bad, even if his carelessness had nearly killed them more than once and his inability to hold a job had meant he was constantly broke and she often had to help him out. "We were such good friends," she remembered. "We really got on well together; we were pals, We had great fun together."[31] But what was Jessie to do? She couldn't go on like this forever, she reasoned. The romance between her and Bill had been over for some time. "I had fallen out of love with Bill Lancaster ages before," she said later. As far as she was concerned, it was past time that she had her own life.[32]

By the end of March, Bill was starting to get the feeling something was going on back home without him. After making a forced landing in a pass between mountains and sleeping in a small town nearby overnight, he and his future employers had arrived in El Paso, but there was no lovelorn correspondence awaiting him. "No news from Chubbie!" Bill exploded in his diary. "She has disappointed me far more than this damned expedition." When he did receive a letter three days later, he complained that it read like Jessie had "just dashed off a note as a sort of duty." He hoped that Haden was "keeping his promise to me, feel sure he is. But Chubbie—Hell." Right now, though, Jessie was the least of Bill's problems. Russell finally revealed to Bill the shady plans by which Latin American Airways was really going to turn a profit, and Bill found himself caught in a bind of his own making. Unable to admit to Jessie that he'd been naïve all over again, Bill wrote instead to Haden about his problem and lack of trust in Russell, who was, he claimed, trying to get him to take a few Chinamen over the border. "You know I am devoted to Chubbie," he wrote. "For her I would do anything or risk anything.... Now we are within reach of some cash if I like to take a little chance and make a little trip with an unmentionable cargo." Justifying his thinking by pointing out that Shelton was likely to go ahead and that Russell seemed to have thought the operation out well, he then attempted to gain Haden's collaboration "because of our newly-formed friendship" by asking him to sound Jessie out about whether she'd be likely to agree "without fully disclosing the hand." He then attempted to justify his participation in what he clearly now knew to be an illegal business by invoking the name of love. "For me, life would be very empty if I lost Chubbie," he wept. "To me, she is the all important thing. If I thought my return in a penniless condition meant losing the kid, I would take even greater chances to return with money than might be prudent."[33]

On the other hand, Bill wrote to Jessie that he was taking no chances and that he'd return if he and Shelton couldn't find a legitimate means of making money. Jessie immediately replied, advising him not to get involved in smuggling. "You are in a dog house on a chain if you ignore my advice," she cautioned him. "This is no time to take chances." They were a "damn slippery crowd," she observed. Bill agreed, writing back that he didn't care if he returned without money, as long as she stuck by him. Haden also wrote, advising Bill not to get involved, reminding him he'd risk deportation and the loss of his reputation. Bill replied to Jessie that he'd eaten only one thirty-cent meal in the last couple of days and hadn't been able to wash his clothes. He added unknowingly, "Thank Haden for his kindness and everything he's done for you. Somehow I felt I could trust him more than anybody in the world. I hope this is true. All my love, little sweetheart; I long to give you a big hug and feel you close." Once again, however, Bill couldn't make up his mind what to do, and so he stayed with the group. He changed his attitude about Jessie again, too. When they finally reached Nogales, Arizona, at the end of March, Bill complained in his diary that there was "no news from Chubbie. Why does the child not write?" He was more judgmental the next day when he still hadn't heard from her: "The little devil; I should not think of her so much. She doesn't deserve it. Hell!"[34]

Meanwhile, Jessie was enduring some emotional hell of her own; neither she nor Haden could decide what to tell Bill. But falling in love can be hard to hide, and in the end someone else told him. On 2 April, when Russell received two letters from his wife, Bill finally found out why he'd not heard from Jessie. For reasons of his own, possibly because Bill had been going on about his perfect love, Russell decided he should share his wife's observations about what was really happening back in Miami. "Chubbie and Clarke came around tonight," she had written in one letter. "I really think that Clarke has gained Chubbie's affections and Bill has lost them. There is no need for Bill to return as far as she is concerned. She is madly in love with Haden Clarke and is worried how to break the news to Bill." Not content with first blood, Mrs. Russell then worked the knife in deeper. "I was around at Chubbie's tonight," she continued in the other letter. "She and Clarke got all ginned up together. Don't tell Bill but I believe she is <u>well satisfied</u>." Two underlined words said it all.[35]

Understandably, Bill threatened to leave immediately for Miami, but somehow Russell persuaded him to wait on the chance that he could make some money for his flight. Perhaps Bill reasoned he might need it. The kind of friend Bill could really have done without, Russell told him that he might as well accept Jessie was a lost cause; she and Haden had visited his wife frequently, he revealed, and Haden now "had the inside track." Bill asked Russell if he thought Haden had double-crossed him and Russell assured him Haden had certainly done just that. He would later claim in sworn testimony that Bill then turned on his heel and muttered, "I'll get rid of him."[36] That night Bill would write in his journal, "Mental agony. Hell!!"[37]

Bill promptly wired Jessie every cent he had and for the next two days he tried to call her but without success. Finally, on 5 April, he did receive some letters from her, but they were unclear as to her feelings, despite his desperate attempts to read between the lines. "Much disturbed," he diarized. "Ill with worry. Chubbie darling, what is it all about?" He lamented that she must need money, but he had absolutely none he could send her. Desperately, Bill begged some money and wired it to her, but her brief reply only worried him further. "If only she would say something more, such as 'Don't worry, I still love you,'" he fretted in the words of rejected lovers everywhere. He resolved that upon his return they would "have it out" once and for all so he wouldn't have to go through this anguish again. "I adore her and want to see her happy," he wrote. "If only she didn't drink while I was away I would feel OK. She has two personalities. Chubbie drunk and Chubbie sober. I love both personalities. Is HC trustworthy is my problem. Chubbie angel, it won't be long before I see you and hold you."[38]

It wasn't going to be as easy as that, though; it never is. When Jessie finally picked up the phone the following day, Bill was left wishing she hadn't. Instead of empathizing with his emotions, she firmly gave him a piece of her mind. She was sick of being left alone and penniless, she declared, hinting strongly at what Bill already suspected: she'd received a better offer. Within the next couple of days, the other shoe fell in the form of two letters: the first was from Haden and the other from Jessie. They both expressed their regret but confirmed Bill's worst fears: they were in love and had decided to marry. "I hate to say I told you so," Jessie wrote, "but it is exactly what I thought would happen.... At present I have 40 cents and Haden is broke. We are living on canned soup, bread and jam.... I told you when you cleaned me out of the little reserve I had in the bank that I would be stranded in the end. I have used every cent I have in the bank and am penniless. I am feeling quite desperate."[39]

Haden wrote that they had fought against their feelings but were powerless in their love for each other:

> We tried to talk each other out of it.... I know its [sic] going to be a blow for you, old boy, but I am faced with the obvious choice of hell for one of us and heaven for two of us, or certain hell for two of us and not much better for the other one. You have said your one aim was for Chubbie's happiness. I don't know that my word means much now but I give it that I will always do everything to make her the happiest girl in the world. If I fail in this I stand ready to answer to you. I beg you to think it over sanely. Try and see your way clear to helping us get it over smoothly. If you lose your head, I'm afraid I can do nothing, but I will meet you halfway. It would break Chubbie's heart if we were violent. The decision rests with you.[40]

To his credit, Bill did not reply harshly, writing instead that "I am no dog in the manger. Hold your horses until I arrive. I insist on being best man, and will be the best friend to you both for life. Your happiness is my happiness."[41] Privately, though, Bill panicked. "I'm going to Miami whatever happens," he wrote in his diary, "to find out first hand all about everything. I love you Chubbie. I have done my best—but failed. Want to talk things over with you my sweet."[42] Three days later, on 20 April, Bill landed in Miami carrying a loaded revolver in his bag.

On the way home from the airfield, Bill bought cigarettes and groceries and paid for Haden's laundry. Dinner that night was endured in an atmosphere strained with artificial politeness. Then Bill took the first shot. "Well," he asked, "what is this nonsense you tell me?"

Jessie and Haden protested it wasn't nonsense at all; they were going to get married. Bill turned to Haden. "You've let me down, Haden," he snorted. And so the argument began. At one point, Bill made the rather radical suggestion that if Jessie and Haden were so short of money he'd crash his plane so they could collect on his $1,000 life insurance policy, no small sum in those Depression days.[43] He then suggested rather more sensibly that perhaps they should wait a month before marrying because he wanted Jessie to be sure of what she was doing and, besides, they should also make sure that Haden's divorce had been finalized. Haden bristled at Bill's blatant paternalism, feeling his alpha male position slipping from under his feet. Leaping to his feet so abruptly that his chair hit the wall, he protested: "I don't want you to break down my wagon!"

"I don't want to break down your wagon," Lancaster reasoned. "I just want her to be sure." Then he got a tad pushy: "I swear before God, if you don't make her happy I'll come and take her from you." Trapped between the two people for whom she cared most in the world, Jessie uncharacteristically broke down in tears. Bill tried to comfort her, but Haden loudly protested that was now his role. At that, Jessie ran out of the room into the kitchen. Bill quickly followed, automatically putting his arms around her to comfort her. Quite naturally, Haden became upset, accusing them of laughing at him behind his back. He refused to leave Jessie alone with Bill. Shaking his head, Bill went upstairs to read his mail. Jessie sank onto the sofa in the living room and Haden slumped down alongside her and put his head in his hands.

"I can't stand it any longer," he wept. "I'm going crazy."

Leaning back against him, Jessie felt completely overwhelmed by a situation spinning more out of control by the minute. "Haden," she sighed sadly. "If we could only just end it all together."

Haden put an arm around her and replied, "Yes, if only we could."

Jessie's despair was only momentary, though; the next minute she brightened up and suggested they could always go to the West Indies and set up a commercial air business. But Haden didn't recover from emotional troughs quite as quickly, and he felt too depressed to even think about it. Eventually, they all wound up in Jessie's bedroom, Bill at the foot of the bed and Haden relegated to a stool by the dresser, while Bill tried to steer the conversation away from personal matters by talking about his recent trip. Nevertheless, Haden still felt threatened, all too aware that Bill and Jessie had years of flying experience in common about which to reminisce. He, on the other hand, had none of that background at all and so he was soon accusing Bill of using that in an attempt to win her back. Bill just repeated that he'd be happy for them to marry in a month, and then he sensibly proposed that he move out of the house the next day.[44]

At that point, a completely exhausted Jessie suggested they all sleep on it. Haden warned her to lock her door in case Bill tried anything during the night such as talking her out of the wedding, and then he and Bill retired to the screened porch and talked.⁴⁵ Jessie later said that, after going to bed in her room she began to read a detective story, hearing the murmur of the men's voices and Haden's laughter from down the hall. Finally, feeling reassured everything would be all right in the end, she turned out the light and fell asleep.

It seemed like only a minute later that a furious loud banging on her bedroom door awoke her, but she saw it was about two-thirty in the morning. Stumbling out of bed, she opened her door to a frantic Bill.

"An awful thing has happened," he cried. "Haden has shot himself!"⁴⁶

Ten

An Awful Thing Has Happened

"Don't be ridiculous!" Jessie recoiled. "There isn't a gun in the house."[1]
She thought at first Bill must be just joking around.

"Yes, there is," Bill replied to her horror. "It's the revolver I brought back to replace Huston's, which I had to leave in Texas."

"What revolver?" Jessie cried.

Bill started to explain to her about buying the gun and ammunition on the way home from St. Louis to replace one belonging to Tancrel's Miami attorney, Ernest Huston, that he'd had to pawn during their trip, but Jessie didn't wait to hear the story. Pushing Bill out of her way, she ran into the sleeping porch, where she found Haden lying in his bed on his right side, face covered in blood from a wound in his temple. Fetching water and a cloth, she bathed his face, holding his head, crying and begging him to speak to her. He only moaned and twitched feebly. She added a cold compress to try to stop the bleeding.

While Jessie was trying to help Haden, Bill grabbed the telephone and called the emergency operator for a doctor, saying only that a man had shot himself. When he returned, Jessie was still wiping blood from Haden's face. They waited, but no doctor arrived and so Bill called the operator again. While he was on the telephone, Haden moved slightly and Jessie noticed a revolver that had slipped down underneath his right side. She reached to move it, but Bill quickly put the phone down and said: "Don't touch it! You don't want your fingerprints on it. Leave everything as it is."

Still no one came, so in desperation Jessie called Mrs. Clarke's physician, Dr. Carlton Deederer, who assured her he'd be there soon. But minutes dragged by and Dr. Deederer didn't show up, either, so Jessie called Ernest Huston, attorney and secretary for Latin American Airways, the only lawyer Jessie and Bill knew. Haden had shot himself, she sobbed. Huston replied he'd be right over and was there in fifteen minutes. In the meantime, Jessie had called Dr. Deederer again, who explained that he'd become lost and returned home. Furious at the delay, she made sure he had explicit directions this time and implored him to hurry. While she was talking, Huston drove up and rushed in the door. Bill then showed him two suicide notes ostensibly written by Haden. Slamming down the phone, Jessie queried Huston what they should do with them, because she feared a scandal if they were discovered. Bill suggested tearing them up, but Huston cautioned against doing that as he already felt that the notes would prove important in any future debate about Haden's death.

Going upstairs, Huston found Haden still lying on his bed in a pool of blood, unconscious but clearly alive. Studying the scene with a lawyer's careful eyes, Huston, too, noticed the gun partially concealed under Haden's body. Bill foolishly asked him if it would be all right to say it was Huston's gun, insisting that he'd bought it for him, but a shrewd Huston was already distancing himself from this blood-soaked mess and gave him a flat, "No," repeating it when Bill made another suggestion: that they should say the gun belonged to Latin American Airways.

Suddenly, an ambulance pulled up at the gate, but the only person in it was the driver, Charles Ditsler. "Where's the doctor?" Jessie cried. No one had told him to bring one, Ditsler replied. Dr. Deederer, the one man they now really needed, had still not arrived and Jessie protested that Haden shouldn't be moved until the doctor had seen him. Bill asked Ditsler if he thought Clarke would be able to talk again, but Ditsler, having seen a few head wounds in his time, was certain that Haden wouldn't be saying anything again to anybody in this world. If they didn't want him to die in the house, he added bluntly, then they'd better get him to a hospital—fast! But Jessie, still unsure about moving Haden without Dr. Deederer's opinion, still hesitated and so a frustrated Ditsler called the ambulance's owner, local funeral director O.C. Yeargin, who drove out to the house himself, about thirty minutes after Huston had arrived.

Once there, Yeargin took one look at Haden and agreed they needed to leave for the hospital immediately, and Jessie finally conceded. By then struggling to breathe, Haden was quickly stretchered out the door and into the ambulance for the journey to hospital, accompanied by Huston and Bill. Huston would later recall that on the way Bill said to him, "I wish to God that Haden could speak so he could tell them why he did it." Jessie waited for Dr. Deederer, who finally found their location a few minutes after the ambulance departed, and together they left for the hospital. Once again, however, Deederer became lost in the dark, and it was some time before they got there. The conversation in that car must have been a very interesting one.

Waiting for them at the hospital were two police officers, one of whom was Officer Earl Hudson, who by a strange coincidence, it would later be

Ten. An Awful Thing Has Happened 173

revealed, was married to the sister of a certain Peggy Brown who would later claim she was also engaged to Haden. The officers returned with Jessie to the house to "examine" the scene of the shooting. Although fingerprints were being used by then in the U.S. as a means of identifying criminals, Hudson was seen by Jessie to wrap the blood-coated gun in his handkerchief and stuff it in his pocket, obligingly rubbing off any fingerprints that might have been on it, although he later attempted to convince a laughing courtroom that he'd put the gun into a cardboard box before thrusting it into a pocket where it clearly would never have fitted. "What a dumb, thick-headed cop," the defense counsel declared. Described as an "emergency police officer," Hudson may not have even been a member of the regular force and certainly didn't act like anyone with respect to a crime scene. He then asked Jessie if Haden had left a note. Although Bill had asked her to destroy the two suicide notes he'd shown her, Jessie hadn't wanted to do that and so she handed them to Hudson. They then went to tell Mrs. Clarke what had happened, after which they all returned to the hospital. Hudson would later testify that he was also asked by Bill if he thought there was any chance Haden would become conscious long enough to talk. But Haden never did regain consciousness, dying at eleven that morning of 21 April without shedding any light on the mystery of his death. Jessie didn't know until she was told she had to identify the body. By now, Jessie had called a friend who was a lawyer, and the two of them went with Ida Clarke to identify Haden at the mortuary. The lives of Jessie and Bill would never be the same again.

From the beginning, State Attorney N. Vernon Hawthorne, aged forty-two and five years on the job, with a deep voice and a deceptively paternal, considerate manner, never seriously considered that Jessie played any active role in that night's events, although he did think she might have been a motivation for it all. However, he had a very different opinion about Bill Lancaster and so, with Haden dead, Hawthorne promptly sent two detectives to the hospital to bring Jessie and Bill in for questioning to the Dade County Courthouse. The city of Miami lies within Dade County, known as Miami-Dade since 1997. It's the most populous of Florida's counties, and its impressive 361-foot-high courthouse, at the time of construction the tallest building in Florida, reflected that significance. Completed in 1928 to a design by A. Ten Eyck Brown, the building's twenty-eight floors contained not only the court, as well as legal and administration offices, but also the county jail, which, in a rather novel "maximum security" design feature, occupied the top nine floors.

Jessie and Bill were asked to surrender their house keys, and their initial statements were taken. It was only then that a dazed Jessie realized for the first time that they might actually be under suspicion.[2] They were then separated and accommodated in the cells on the nineteenth floor. There Jessie was kept alone for some hours in a small cell, with only her thoughts for company. "They were desperate thoughts," she wrote later. "I realized that I was a woman, alone, ten thousand miles from my own country."[3] Only after some considerable time had elapsed was she finally escorted to Hawthorne's office to answer some serious questions, a daunting experience for someone who'd never previously been exposed to interrogation by any legal system, let alone one in a foreign country. Jessie was grilled under lights during the night for four or five hours while she sat in a chair surrounded by five interrogators in a semi-circle who fired questions at her one after the other with machine-gun rapidity. Afterwards, she was returned to her cell that was alive with cockroaches and mice, where she was given bad-tasting water and inedible food.[4]

Despite the need for an autopsy to provide answers to some obvious questions, only a superficial examination of the outside of Haden Clarke's body was carried out by Dr. Deederer. This was the only examination carried out on the body before it was interred and evidence of such damage as skin bruising was subsequently lost. So, in the early evening of that Thursday, Jessie was temporarily released from custody to attend Haden's funeral, which took place only a few hours after his death, and she and Ida Clarke wept and embraced during the service.

When Jessie returned to the courthouse, "broken by grief, her eyes red from weeping," she had a few words to say to waiting reporters. "It was not a love triangle," she declared, denying the story already being pitched by the press. "I was alone in the world, not having a place to go or anywhere to rest. I loved Haden Clarke. Captain Lancaster and I were merely pals. We were never in love. I realized that I loved Clarke while we were collaborating in writing my flying adventures."[5] Then she reentered the courthouse, where her interrogation was resumed and where she and Lancaster were detained in custody while their house and the airplane were ransacked by the police. Meanwhile, Ida Clyde Clarke was also having a few words to say in her own inimitable fashion, prompted by doubts over the authenticity of the suicide notes. She no longer believed her son had taken his own life, she declared, claiming that he had disliked guns and killing so much that he didn't even play with toy guns. She also claimed, perhaps attempting rather clumsily to protect Jessie, that she knew nothing of any relationship Haden might have had with her, although she was certainly aware of his friendship with Bill.[6]

On Saturday, 23 April, Hawthorne announced at a press conference, which Mrs. Clarke, Dr. Deederer and investigator J.B. Rowland attended, that he, too, had doubts about the suicide theory. He wasn't satisfied with the origin of the two suicide notes, either, he declared, especially now that Mrs. Clarke was convinced the signatures and word spellings were not those of her son. Handwriting experts had backed her claims and so further analysis of typewriter samples would be carried out by Hawthorne's expert, who bore the wonderful name of Fred Flake, but investigators were already noticing similarities between Lancaster's writing style and the notes. Deederer duly announced that he did not find any trace of powder burns on Haden's face nor did his skin or hair show any traces of scorching that would be the result of a blast from the barrel of a gun held to the head at close range, but he had found "several bruises behind the right ear that might have been made by blows from a pistol butt or a fist." He also reported a skull fracture that was not, in his opinion, a result of a bullet penetrating the skull. The mortician who had taken charge of the body reported noticing bruises on Clarke's right shoulder.[7] Apparently, however, there had also been another interesting revelation: a young lady named Peggy Brown had visited Hawthorne's office earlier that day to claim that she, rather than Jessie, was Haden's true fiancée and to volunteer her help in determining whether Haden had committed suicide or not.[8] At the end of a very busy day, Hawthorne evidently decided that Jessie had been interrogated enough for now. As far as he was concerned, her answers had confirmed his original theory that she was not actively involved in the incident. Finally released from custody, Jessie went directly to the Clarke home, where she was warmly welcomed by Mrs. Clarke and their lawyer, James Lathero. Long into that night, the trio discussed possible legal strategies.

Despite being unable at this point to answer the question of whether Haden's death was murder or suicide, Hawthorne also decided to release Lancaster. Nevertheless, Hawthorne was already aware that investigators searching Jessie and Bill's home had found not only Lancaster's personal diaries but also letters and other documents associated with the somewhat shady intentions of Latin American Airways that revealed the operators' plans to engage in smuggling drugs and people across the Mexican border into the U.S. So, waiting right outside Hawthorne's office were Federal officers who now wanted to ask Bill a few questions about those matters, and Bill was taken into custody once more. During the ensuing interrogation, though, the persistently virtuous Bill remained adamant that the moment he'd learned of Latin American's plans to smuggle Chinese and their gold over the border he'd withdrawn from his contract with them. He apparently didn't think it necessary to clarify that in fact it had been Jessie's letter that had sent him hurrying home rather than his conscience. The next morning, Hawthorne called another press conference. Like all astute politicians, he couldn't resist the opportunity to reveal that only he had been allowed access to certain information, because he was worthy of that privilege. Reminding reporters that had he decided to charge Lancaster, Hawthorne's name would be on the front page of every newspaper, he added, "It has been my privilege to see into the depths of a man's soul through his private diary, which was never intended for any one's eyes but his own, and in all my experience—which has been broad–I have never met a more honorable man than Captain Lancaster. Because of this, I'm going to release him."[9]

Walking out of the courthouse doors into the morning of 24 April, Lancaster followed Jessie to Mrs. Clarke's door, trailed by reporters. In full view of gathered media, and perhaps for their benefit, he placed his hands on her shoulders, looked into her eyes and declared: "I want you to know that I didn't kill your son." Calmly looking back, Mrs. Clarke replied, "Captain, I'm glad to hear it."[10] Then, demonstrating either a lot of grit or a very clear conscience, Bill returned alone to the same house in which Haden had died, while Jessie remained with friends of Mrs. Clarke. "I am still exhausted from the ordeal. Everything has now changed," Jessie informed curious reporters. She planned to stay in Miami and complete her book during the next few months, with the help of Ida Clarke, but it would be more serious in tone now. Although returning to flying "may give mental ease," she thought, there was still no money there to finance any further record attempts.[11]

During those long hours sitting alone in his cell, Bill had time to think long and hard about those "suicide notes," and now at last he came clean to Jessie about them. After all, he'd lied to the police about the origins of material that was now evidence in a possible murder investigation, and so what had started out as a thoughtless act born of panic could well be interpreted by Hawthorne as perverting the course of justice. Bill knew that once he revealed to Hawthorne he'd written those notes, then he was going to need Jessie's support and absolute faith in his innocence if he was to have any chance of escaping the inevitable murder investigation. So, being Bill, he of course told Jessie that he'd faked the notes to protect her. An honorable knight of the air like Bill couldn't possibly have attempted to deceive officers of the law in order to save himself. No, he'd written those notes to save the honor of his beloved "Chubbie." Cleverly playing on Jessie's guilt that these events had been brought on by her behavior with Haden, Bill insinuated that they were now in this together, that he had in essence broken the law and been prepared to accept the consequences for her.[12] So, now Jessie owed him a debt of honor and also shared some of the responsibility. If the police had thought he'd shot Haden, Bill hinted, then he would not have been able to bear Jessie's pain or see her good name dragged through the mud. Instead, he'd reasoned in those few chaotic minutes, if there was evidence confirming Haden had shot himself then the case would be quickly closed and it would all just quietly go away. Of course, Lancaster could then be Jessie's comforter with her all to himself again, but that was his secret knowledge. Jessie, who'd thought Bill's behavior about the notes a little odd but hadn't for a moment considered he'd created them himself, couldn't believe what she was hearing. Bill's thoughtlessness had nearly got them killed before, but not executed![13]

She quickly called in their lawyer, James Lathero, for advice, but when he also said to keep quiet and not say anything about it Jessie reminded him and Bill that Hawthorne already had handwriting experts at work analyzing the notes and it was probably going to be a close race between Bill's confession and the experts' verdict that the notes were fakes. Bill's character would look much cleaner, she pointed out, if he was pre-emptive and pleaded to Hawthorne that it was all just an honorable mistake, rather than wait for Hawthorne to find out for himself and then arrest Bill. With Jessie pushing Bill all the way, one version of the story says, they immediately returned to the courthouse and laid all their cards out on the table for Hawthorne to see in an interview that was recorded by the State Attorney's stenographer.[14] However, many years later, Jessie recalled a slightly different version of events in which Bill was a complete coward and flatly refused to go with her and so she had to face Hawthorne alone. Bill had written the note in a panic, she said, because he thought he'd immediately be accused of the murder. "I felt," she remembered, "that had I not told them the truth it would have meant a lot of questions and suspicions and would leave a smear if it had to be dragged out of us under interrogation."[15] Whoever went to the interview, they were only just in time, for there on his desk was the report from his document expert, J.V. Haring, the same man who'd been called in to examine the Lindbergh kidnapping ransom note, that confirmed Bill's confession the notes were forgeries. Hawthorne appeared more relieved than surprised; he'd had suspicions about the validity of the notes all along. When they left his office, though, Hawthorne left them with no illusions about the eventual outcome. He'd just been presented with a gift-wrapped suspect, and he didn't waste any time unwrapping. A few days later, on 2 May, Lancaster was formerly arrested and reincarcerated in another of the Dade County Courthouse upper-floor cells, while Hawthorne and his team set about gathering the evidence to prosecute their case. As far as they were could see, it was not going to be a difficult one.

When Hawthorne announced Lancaster's arrest to the waiting reporters, Jessie immediately ran her colors up the flagpole, declaring in her own press conference that she was "absolutely confident that everything will come out all right. I know the truth will be learned and that Captain Lancaster will be cleared. He is innocent and I know it. My faith in him remains unshaken. It has never wavered in the past, or now, and it never will."[16] Well, that wasn't entirely true, of course; her faith in him had wavered so significantly that she'd agreed to marry someone else, but she was talking to the media now. Jessie was attempting to make up for damage done by remaining steadfastly at her friend's side in his time of trouble.

Neither Jessie nor Lancaster showed any inclination to hide from the press during the following weeks, especially Lancaster. Privacy does not seem to have been an issue while the trial was in progress, whatever Jessie might have said later about the way it was conducted. Lancaster would even be reprimanded by the judge at one point for attempting to hold a press conference in the courtroom. It was typical Lancaster, then, to hold a meeting with reporters in the courthouse elevator on the way up to his cell, telling them that "I am absolutely innocent and I know that the outcome will prove this. Mrs. Keith Miller is a devoted friend and she, at the right and proper time, will tell the truth about the death of Haden Clarke as far as she knows."[17] No pressure for Jessie there. Not that she could really tell "the truth" about Haden's death, anyway; she hadn't been in the room when he was shot.

Hawthorne was just as capable at using the press as a means to his end, though, and while a grand jury convened to decide whether Lancaster should be indicted, someone in Hawthorne's office conveniently leaked to reporters most of the pertinent entries from Lancaster's personal diaries, particularly the ones containing his feelings for his "darling Chubbie." As could be expected, the press had a field day with lines such as, "Chubbie, angel, it won't be long before I see you and hold you," along with his revelations about attempting to take the blame for that auto accident when Jessie was driving drunk and then, when given that minor penalty, crowing triumphantly that "American justice is all wet." Naturally, Hawthorne expressed suitable righteous indignation that the leak had occurred and the grand jury dutifully subpoenaed and questioned six reporters, but in the end no conclusions were publicly announced about how selections from those private diaries that were handed personally to Hawthorne came to be passed on to reporters.

Jessie, who up until now had rarely spoken in public about her emotions, must have been mortified to see Lancaster's feelings about her displayed for everyone to read. Nevertheless, she visited him regularly and he kept two pictures of her prominently on view in his eight-by-ten-foot cell in which he was to remain for the next three months, whiling away the time reading three to four books a day and listening to the radio. The food was so bad that Jessie took something to the jail for him to eat every day, although she was only allowed to actually visit him twice a week for thirty minutes. As well, Jessie also become her own private detective in order to hunt down likely witnesses for her friend's defense, and soon the stress took its toll. Though she was never a big person at any time, Jessie's weight dropped to an alarming eighty-five pounds.[18]

While telling reporters he was confident that twelve reasonable men could be convinced of his innocence in Haden's death, Lancaster certainly did not have the money to hire the legal counsel to prove it. Nor did Jessie for that matter, and so she set about trying to raise the funds. It wasn't easy: there were only three replies to the fifty telegrams she sent to people she knew asking for help. In the end, their defense was mainly funded by Lan-

Ten. An Awful Thing Has Happened

caster's eighty-year-old father, while Jessie's mother regularly contributed towards living expenses. Meanwhile, Jessie discussed with Hapgood and Lathero their options of good defense counsel, eventually deciding that the best person for the job would be prominent South Florida attorney James Carson. But, the two men warned, he was a very busy man who might not be able to take the case and, even if he did accept it, he'd demand a hefty fee. Jessie was not to be dissuaded, however, and so Lathero arranged a meeting with the great man. Born into a Kissimmee, Florida, cattle ranching family who had been in Florida over a hundred years, the formidable broad-shouldered, six-foot-tall James Milton Carson had been in

James Carson, Jessie and Bill's defense lawyer (courtesy of Crisanda Singer).

practice since 1916. He was also a member of the law faculty of Miami University and an authority on Florida history. Jessie would later recall that on meeting Carson she first thought he was rather a big, terrifying man to look at, with creases in his full face. When she asked him if he would defend Lancaster, he reacted, "I wouldn't touch it; he's as guilty as hell!" The feisty Jessie indignantly replied that she knew he was wrong, that if only he went to see Lancaster personally he'd see that he wasn't the kind of man who could have murdered Haden. "He's not a coward," she concluded emphatically.[19]

Impressed by Jessie's earnestness, Carson gave the matter some thought for a day or two. Like Hawthorne, he was clearly aware of the amount of publicity this trial could generate, locally and internationally, and so he eventually took himself off to the upper floors of the courthouse to talk to Lancaster for several hours. Just as Jessie had predicted, after meeting him in person and listening to him Carson was so convinced by his demeanor, his apparent honesty and his active service during World War I that on 5 May he agreed to take the case, assisted by Lathero. Then he sat down with Jessie and had a serious conversation about their chances. It was not going to be easy, he warned her. Bill's character had been damaged by his forging of the notes, and he was foreign. They would have to make him look better by making

Jessie look worse. "He warned me," she wrote later, "that my life would be painted in the most lurid of colors; private incidents would be 'whooped up.'" However, if Bill was to emerge from it all as Sir Lancelot, Jessie was going to be Joan of Arc: the martyr for her cause. "I was free," she wrote later. "I could go anywhere, any time. [Or] If I wished I could stay and have my name dragged through the mud of publicity, and see the small place I had made for myself in the aviation world lost, perhaps forever. I stayed."[20]

James Carson with his wife, Ruby Leach Carson, along with children Jack and Carol (courtesy of Crisanda Singer).

Two days later, after consulting further with Dr. Deederer, Ida Clarke publicly declared that she believed her son had been beaten and then murdered, but she stopped short of naming Bill as the murderer. While previously she'd not commented about the lack of an autopsy that might have produced evidence to support her claims, now his mother suddenly demanded that Haden's body be exhumed and examined. She claimed to be "perfectly attuned with the after-life personality of her son," and that she would not be surprised to receive a spirit message from him because, she revealed, Haden had been a disciple of Spiritualism just like his mother and believed, "even more strongly than I do, that spiritual communications are possible between strong personalities who are mental affinities. We believed in a life after death."[21]

Finally, to no one's surprise, the grand jury declared that there was a case to be heard, and a tired and haggard Bill Lancaster was arraigned on 17 May at the Miami Circuit Court before Judge H.F. Atkinson, where he pleaded not guilty to the charge of murdering Haden Clarke. After some discussion, the judge set the trial date for 5 July in order to allow time for Carson to fly in character witnesses from out of state. A few days later, Atkinson granted Mrs. Clarke her wish for an exhumation after Carson outlined a series of reasons for the necessity for a detailed examination of the remains that included the surprising lack of an autopsy in the first place, that there was no eyewitness to the shooting and thus the body itself was the only source of evidence apart from the accused, and that bruising had been reported on the body by both Deederer and the undertaker. A panel of four Miami doctors was convened as a commission and Haden's body was promptly disinterred and examined. As Jessie had anticipated, the unanimous verdict of the panel was that the wound had been self-inflicted. Interestingly, when the coroner was later criticized by Carson during the trial for not initiating an inquest into Haden's death, which would have required an autopsy, he claimed he'd received specific instructions from the State Attorney's office at the time not to hold one.[22]

However, by the time Haden's body was being examined, Jessie had legal problems of her own. To the surprise of even the American Consulate back in Australia, she was arrested by immigration authorities on 22 June and charged with illegal entry into the U.S. Considering the number of years Jessie had been in the country and the number of times she had traveled back and forth across U.S. state and international borders without any problems, this was without doubt a setup and Jessie knew it. When arrested, she was merely informed that some unidentified person had lodged a complaint, so not only was the Department of Labor using that well-known "anonymous tip" as reason for arrest, but they had obviously waited until Carson, who could have immediately posted a bond, was out of town. When Jessie demonstrated proof that her initial entry into the U.S. from Australia had been approved by the appropriate authorities at the time and was quite legal, the Department then switched stories and endeavored to charge her with illegally entering the U.S. from Nassau the previous year. Jessie found herself lodged once again in the women's division of the courthouse cells, this time on the floor above Lancaster. "Evidently I am insufficiently harassed and needed another worry," she commented wryly.[23] Needless to say, Carson was considerably annoyed when he returned to discover what had happened, and he quickly posted a $1,000 bond to have her released from jail while they sorted matters out. An immigration hearing date was penciled in for the beginning of August, but Lancaster's trial date intervened and any hearing was consequently postponed until after that event.

Before the trial began, Jessie gave an uncharacteristically revealing and quite poignant press interview about the nature of her relationships with Lancaster and with Haden, who, it was becoming all too apparent, had not been entirely truthful to her about his life. "My love for Captain Lancaster was worn out before I met Haden," she said,

> though I consider Lancaster the finest human being. Haden and I suddenly loved with the maddest rapture, hysterically. He came and helped me with my autobiography. We hardly worked; we loved madly and insanely. Both of us were broke. Haden always said, "God will provide." I usually did. I could scarcely buy food, and the lights were often out. We were delighted in each other and hated separation. When downtown for a few hours, Haden would spend a precious nickel and ask if I still loved him. He adored my voice, and I read aloud every day, mainly from Oscar Wilde's "The Ballad of Reading Gaol," which was like a symbol of our love. In great sorrow, I am realising the great wrong Haden did me. I planned to marry him. Captain Lancaster, whom I loved without being in love after years of companionship in flying adventures, was upset but only wanted my happiness. On the dreadful night I heard Captain Lancaster and Haden talking and laughing. Haden's voice and laugh were the last I remember of him.[24]

Shortly before nine-thirty on the morning of 2 August, Lancaster's trial opened in the courtroom on the sixth floor of the Dade County Courthouse before Judge Atkinson. A crowd estimated at some 2,000 had been gathering in the heat outside the courthouse since six in the morning, some carrying their own brightly colored cushions for the hard court benches. As soon as the doors opened, the crowd stampeded into the building and headed for the courtroom. So many people attempted to fit into the room, which could only hold 200, they jammed the stairs all the way down and out onto the sidewalk in front of the building. Women were trampled and several fainted in the intense heat. Although this was the middle of summer, the courtroom would remain packed with spectators and reporters for the duration of a trial that would be covered by newspapers throughout America, Britain and Australia. Reporters had been granted special permission to take flash pictures within the courtroom, something that never happened in an Australian court. To

Jessie, the courtroom seemed hectic, where "everybody was excessively friendly, the prosecution and the defense laughing and joking together. It seemed to me more like a stage comedy than a trial where a man's life was at stake." Direct line telephones to northern states had been installed in the courthouse for reporters along with noiseless telegraph stations within the courtroom itself. Jessie recalled that the press table was bigger than the judge's bench and the prosecution and defense tables combined and, to her surprise, the judge dressed simply in a lounge suit with no wig. The constant waving of fans, she recalled, gave an onlooker the impression of a room in perpetual motion.[25]

First, Carson moved to be granted a continuance until 15 September due to the illness of one of his key witnesses, Dr. P. Dodge, who had been head of the autopsy panel, but Atkinson quickly overruled the motion. Then, over the next two and a half hours, the twelve men of the jury were selected from the pool of 100 people who had been called. Eventually, a grocer, two merchants, a retiree, two contractors, two salesmen, two real estate dealers, a farmer and a meter checker survived challenges as to whether they were war veterans, had flown airplanes, were prejudiced because Lancaster was English and, importantly, "Can you give this man a fair trial in spite of sordid details that may arise to show that he lived with a woman who was not his wife?" There lay the first clue to the trial's future tone.[26] Another one might have been a frequent newspaper observation that as Lancaster was being photographed in the courtroom before the trial got under way the breeze had fluttered the American flag behind him so that its folds draped over his shoulders.

Once the jury was sworn in, Vernon Hawthorne stood to begin his opening argument for the prosecution, reading the indictment and then painting a graphic picture of the testimony he intended to present. Jessie, dressed in a white silk sports dress with matching hat and yellow shoes, seated at the defense table with Lancaster, listened attentively as Hawthorne declared testimony would show Lancaster loved her and was continually worried about her while he was away, having trusted Haden Clarke to care for her. He had become miserable after leaving Miami, Hawthorne alleged, and upon his discovering Clarke's infidelity he had threatened to kill him. "I'll get rid of him," he'd declared in the hearing of others, "I've seen many dead men; I can see another one." Lancaster had been driven to murder Haden Clarke by jealousy with a pistol he'd bought in St. Louis on his way home, Hawthorne proposed, and then, no doubt to set up in the minds of the jury that Jessie and Bill were people with dubious reputations, he informed them just before he sat down that although Jessie had been exonerated of Clarke's death and was not on trial here, she was nevertheless under bond on charges of illegal entry into the U.S. and that similar charges would be brought against Lancaster once this trial was over.

When James Carson rose to put forward his opening argument for the defense, he first reminded the courtroom that even though Lancaster, war hero and "one of the great and famous fliers of the world," was far from home, he was still being tried under the common law of England, which was also the law of Florida.[27] Lancaster admitted forging the suicide notes, he told them, but otherwise the State had built a case of circumstantial evidence out of a series of suspicious circumstances, whereas the defense's case would be built on the rock of direct testimony. If you began with the presumption of innocence, he argued, to which Lancaster was entitled under the law of both countries, then each of the State's pieces of evidence fell into place to prove Lancaster's innocence, but if you believed him guilty then the evidence would tend to confirm those suspicions. In other words, he clarified, it was the job of the State investigators to look for evidence that would prove Lancaster guilty. It was Carson's job, on the other hand, to identify evidence of his innocence. It was the jury's task to consider if that evidence was true and if it was consistent with the innocence of the defendant.

He then summarized Lancaster's background, emphasizing his father's importance in the UK as a civil engineer and Lancaster's own record as a World War I airman, and reminded the court that a number of character witnesses would appear who had voluntarily traveled great distances to verify Lancaster's reputation. He quickly moved on to the circumstances of Lancaster's first meeting with Jessie, giving some early hints about how he was going to present her part in all of this. She had, he suggested, a "sense for publicity," and a "sort of desire to push herself forward in the world," and so she'd undertaken to raise part of the cost of the England–Australia flight on the condition she be taken along as a passenger because she was not a pilot. This was just a foretaste of what Jessie would have to endure during this trial, and this particular character reference was from the defense! Then Carson told the court that the intimacy of Jessie and Bill's relationship had deepened into "sexual passion" during their flight from England to Australia, reminding the room they were both married at the time, that they had then moved to America and had achieved fame there, that there were common interests that held them together and that it had been their reduced circumstances during the Depression that had brought them to live in Miami. As if Jessie had no status of her own or the ability to make choices for herself, Carson informed the court, "In bringing Mrs Keith-Miller from Australia to the United States with him, although at the time she had a husband in Australia, Captain Lancaster took upon his shoulders a responsibility; and being a man of high sense of honor and high character, which he is, that sense of responsibility has ridden his shoulders ever since."

Although the passion between them may have cooled, Carson pointed out, Lancaster maintained a devotion to Jessie out of pity, love and friendship.

Ten. An Awful Thing Has Happened

In Miami, Carson recounted, Jessie had met Ida Clarke, who had suggested her son could "ghost write" Jessie's book. Haden Clarke, while brilliant, with a vivid imagination and charming personality, was emotionally unstable, "almost unbalanced, probably neurotic and certainly erotic in his disposition." Carson then outlined the story of the days leading up to the shooting, reminding the court that Lancaster had left Jessie in Haden's care, that all parties had been desperately short of money, that letters and telegrams had revealed Jessie and Haden were having an affair, and that Bill had returned to Miami and had bought a revolver on the way to compensate for pawning the one Ernest Huston had loaned him. In front of Jessie and the court, Carson recounted Bill's discovery that Jessie and Haden "had begun having sexual intercourse with each other and that it had continued with excessive violence for a period of two weeks," and that they had only ceased after discovering Haden was suffering from "one of those vile diseases that people who spend their lives in certain ways may expect," insinuating quite clearly that Jessie as well as Haden could have expected it. Carson then outlined the circumstances of that night, offering his moral opinion that Clarke's medical condition should have prevented consideration of immediate marriage, not to mention he'd received a telegram from an existing wife in California informing him that under the terms of their divorce decree he could not marry until January 1933. He proceeded to describe events surrounding the shooting itself, including Lancaster's forging of the suicide notes, a time when he'd "lost his head and done fool things," reminding the jury that Lancaster had been released from custody when he volunteered the information he'd forged the notes and had voluntarily signed the agreement for Clarke's body to be exhumed. Carson warned the jury that he'd be producing photos and exhibits pertaining to the exhumation. He'd begun to point out to the jury the details of how the exhumed remains revealed the gunshot wounds were self-inflicted when Hawthorne reprimanded him for giving evidence for which he was not medically qualified. Carson harrumphed and then modified his wording to "medical testimony will show that" nobody could hold a gun tightly against their head and fire it without leaving powder burns. He then concluded that Haden was a prime candidate for suicide because of his tendency to depression and links with drugs and alcohol. However, in the end each jury member would have to decide for himself "whether the circumstance is consistent with that presumption of innocence which the law affords to this defendant." Jessie and Bill had been desperate at times, he conceded, "but neither he nor she has ever been desperate enough to kill, and there is not a bit of direct evidence that can even remotely connect my client with such a charge in this case." But by now, the room was barely listening; they were all really waiting for the moment when the woman over whom the press was already salivating as the central figure in a love triangle would take the stand and, they hoped, for the sake of salacious gossip and newspaper sales tell all in delicious detail.

They didn't have to long to wait, although some other people preceded her into the stand during the morning of 3 August, the second day of the trial. Ernest Huston, the secretary-treasurer of Latin American Airways as well as their attorney, testified about Jessie's phone call at three in the morning and what had happened after he arrived at the house. At that point, Hawthorne produced a gun that had been admitted into evidence and asked if it was the one Huston had seen at the house, but Huston would only admit that it was "similar" to the gun he'd seen there. On cross-examination by Carson, Huston testified that he had previously loaned a Colt .38 revolver to Lancaster, who later said that he'd lost it. Huston was followed by Hudson, the police officer first on the scene, by Yeargin the funeral director, and Ditsler the ambulance driver, who all recounted their roles in the events of that night.

Then, late in the morning, the State of Florida in the person of State Attorney Vernon Hawthorne finally called Jessie to the stand as a witness for the prosecution for an examination that newspapers judged "soul-searing." Bill was sitting at the defense table directly in front of her, Judge Atkinson was on her right, and the jury sat along the wall to her left. Jessie later recalled that Atkinson had been a very just man, helpful and fatherly when she became upset. "I was shaking with nerves when I was being questioned," she remembered. "I was so frightened in case I put my foot in it. I could only tell the truth as I'm no good at telling lies. It only gets me into deeper and deeper difficulty."[28] Carson had in fact advised her to answer truthfully, as long as she was not long-winded about it, so, in answer to Hawthorne's opening questions, Jessie gave a detailed account of the events leading up to and including the night of the shooting, agreeing that although she had for a while considered the possibility of marriage to Bill, she'd become disillusioned with him after almost five years of waiting to marry. When Hawthorne asked if Jessie had been engaged to Bill, she tartly replied, "You can't be engaged to someone who is still married." So, alone and with no one to turn to after Bill had headed west, she had fallen in love with Haden Clarke.

Jessie testified that when they realized what had happened between them she and Haden had agreed it was the right thing to do to write Bill and let him know the situation and that they'd both read each other's letters before posting them. Later, during the discussion with Bill on the evening prior to Haden's death, they had agreed when Bill suggested postponing the wedding until Clarke's divorce situation had been clarified, and Bill had agreed in turn to move out the next day. Lancaster had seemed saddened by the romance between her and Clarke, she went on, and had even gone as far as offering to kill himself by crashing his plane so that she could collect the $1,000 insurance money as a "wedding present," only abandoning that plan on discovering

A very somber Jessie on the witness stand during the Lancaster trial in Miami, probably on her first appearance, 3 August 1932 (author's collection).

that his insurance company had gone out of business. Jessie's voice broke momentarily as she spoke of locking her bedroom door that night because Haden had told her "he didn't want that *** to come to my room and talk me out of our marriage plans." She hadn't heard the shot that night, she stated. "Do you know who killed Haden Clarke?" Hawthorne lunged. "I am convinced that he killed himself," Jessie parried. By the time she stepped down, a weary Jessie had been on the stand for some six hours. "She is very tired," Hawthorne graciously conceded.[29]

The following day this self-reliant pilot, who was willing to let her international reputation be dragged through the mud in order to help save her best friend from the electric chair, resumed the stand to be cross-examined by Carson. Once again the courtroom was packed at the prospect of dramatic revelations, so crowded that extra chairs had to be brought in. Women still formed a large proportion of the spectators, eagerly leaning forward to catch every word from Jessie's soft voice as she further detailed the events of the night during which Haden died. Already, the trial was attracting international attention; within the courtroom crowd were three British Intelligence officers keeping an eye on developments. Jessie certainly provided the room with revelations as her honest and even-handed testimony laid bare the character deficiencies and mistakes of all concerned, even herself. Contradicting her statement to Hawthorne that she and Haden had not discussed suicide the night of the shooting, Jessie admitted to Carson that indeed she had made "a fool suggestion" to Haden at one low point during the evening about wishing "we could end it all" and Haden had agreed, but that was the only time she had ever heard him mention suicide, although Haden also had periods of despondency when his mother had been nagging him about his drinking bouts. Jessie revealed that Haden had felt inferior in the face of her long-standing friendship with Bill: he'd accused them of laughing at him behind his back and that Bill was attempting to regain her affection. When Lancaster had hugged her at one point that evening, Clarke had told him, "Keep your hands off her," to which Lancaster replied, "I've been listening to her troubles and comforting her for five years. I give you a year to make her happy, otherwise I swear I'll come and take her away from you." Yet everything had been so amicable and friendly when Bill had got off the plane and before they'd gone to bed, Jessie protested sadly. "What's mine is yours, don't be foolish old man," he'd said to Haden.[30]

The next witness for the prosecution was J.F. Russell, one of the partners in the ill-fated Latin American Airways venture, who was already serving a sentence for conspiracy to smuggle people into the country and was consequently accompanied by an attentive deputy sheriff. Russell testified that he'd met Lancaster in El Paso, along with other men involved in the business, to discuss operating a commercial airline between Mexico and the U.S. Bill had been worried about his lack of money, he said, and about the status of his apparently deteriorating relationship with Jessie. Russell, it transpired, had been only too ready to tell Bill that he'd lost Jessie to Haden and to share his wife's letters describing how Haden and Jessie were partying together in Bill's absence. Providing a suitably dramatic coda to the day, Russell claimed under oath that "Bill asked if I thought Haden had double-crossed him and I said I thought he had. Bill turned on his heel and muttered: 'I'll get rid of him.'"[31]

Meanwhile, while Jessie had been exposing her soul before that crowded courtroom, Bill was talking himself up to the media. After repeating his defense that Haden had committed suicide, he declared emotively: "I loved Mrs. Keith Miller; she loved me. Haden Clarke won her love from me but I did not kill him." Bill had been "irked" by his confinement, he lamented, and was now glad the suspense was over and he'd have a chance to tell his story that would be "the whole truth and, therefore, it will include the deep devotion which I have had and now have for Mrs. Jessie Keith Miller, whom I have known and loved for five years." He claimed that after they had made their flight from England to Australia "my love for her was such that our lives have run together ever since." After recounting the circumstances of their meeting with Haden and the events of the night during which he'd been shot, admitting that he'd been foolish to forge the suicide notes, Bill summed up with an emphatic declaration of innocence and that he knew the truth would be brought out by the trial.[32]

Having painted a picture for the prosecution the previous day of a jilted lover out for revenge, Russell faced Carson's tough cross-examination the following morning. Carson set out to target Russell's credibility. Yes, Russell grudgingly admitted, he was currently doing time in the Dade County jail for people-smuggling and, yes, he had previously been in the maximum-security Leavenworth prison for his role in smuggling foreign nationals over the border from El Paso to Fort Worth. Carson pointed out that some of Bill's diary entries revealed he hadn't trusted Russell, who had unsuccessfully tried to persuade him to assist in smuggling Chinese into the U.S., and that despite being completely without money, Bill had written he would not resort to illegal means to obtain it. While the diary entries and letters to which Carson referred also demonstrated Bill's unflinching and passionately expressed conviction that he and Jessie were meant for each other, once again they also revealed that he just as unflinchingly preferred to ignore the fact that Jessie was no longer interested in love with a man who remained a husband and father to a wife and two young daughters back in England.

So, when Hawthorne then took his turn on Saturday to read excerpts from Lancaster's diary written between 1 January and 15 April, he attempted to demonstrate to the jury that it was this same blind conviction and passion that supplied Bill's motive for murdering Haden, namely, "his beloved had been taken by his best friend," of whom Bill's own first impressions had been favorable enough for them to establish that friendship. Lancaster, he declared, was a man in a state of abject anguish and despair as he desperately attempted to swim against the changing tide of Jessie's feelings towards the very man Bill had entrusted in his absence with the care of the woman he loved more than "my very life." Bill had clearly written that "life would be very empty if I lost Chubbie," Hawthorne pointed out, and perhaps it was fear of that emptiness that drove him to shoot the man who would cause it. Bill had written that "I can't stand the strain much longer," and obviously a man in torment, he had scrawled across those fateful pages, "Mental agony. Hell."[33]

That afternoon, Hawthorne continued to add color to the portrait of Bill as a man under pressure by bringing to the stand another member of the Latin American Airways group, Mark Tancrel, who testified he'd twice heard Bill declare that he'd "deal with" Haden. Bill had finally lost his temper in March, Tancrel claimed, in a hotel in Nogales, Arizona, shouting: "I'm through! I'm sick of it. You can paddle your own canoe. I'm finished. I'm going back to get rid of that ***!" Tancrel testified further that it had been he who had heard Bill tell a flying companion in El Paso that "I have seen a lot of dead men, and one more won't make any difference."[34] The next witness was funeral director K.B. Bess, who testified that he'd observed Clarke's right arm had been bruised at the shoulder and that there had been no powder burns surrounding the wounds, although he conceded during Carson's cross-examination that the body hadn't been in the best of condition by the time he saw it. By this time, the jury was visibly feeling the strain of a long week of testimony, and when Bess stepped down they petitioned Judge Atkinson to consider holding night sessions in order to shorten the projected trial period. While he didn't agree with that proposal, Atkinson could see they did need a break and he duly gave permission for the jury to have a supervised recreational outing to Miami Beach and to see a movie.

When the trial reconvened on Monday morning, before a refreshed judge and jury, Jessie was back on the stand. Looking out at the same hot, humid, overcrowded courtroom, Jessie faced the daunting interrogation technique of State Attorney Vernon Hawthorne, who was just as intent on seeing Bill convicted for murder as Jessie was to publicly support his innocence. Although she and Hawthorne might have started out with a grudging respect when they first met, to say they didn't like each other at all by now is an understatement. "Hawthorn turned out to be a swine," Jessie vividly remembered years later.[35] Under his relentless questioning, she had to concede that since his death she'd lost her love for Haden because she'd discovered he'd lied to her about details of his life. Then Hawthorne dropped his first ugly bombshell. Wasn't it true, he asked, that she had become "ill" in Cuba and then communicated that "illness" to Haden? Now, in the cautious public vocabulary of those days, the word "illness" in this context was a clear inference to sexually transmitted disease, and it was a very blatant attempt by Hawthorne to damage Jessie's character. Of course, no such thing had happened at all, and a quietly furious Jessie courageously turned the tables on Hawthorne as she let him know what she thought about his line of questioning:

"You gave your word of honor, Mr. Hawthorne, you would never say anything," she accused, reminding him that after the initial investigation he'd told her, "I've looked into Lancaster's soul and seen an honorable man incapable of shooting Haden Clarke." Haden's physical condition had previously been attributed to her, she declared, and she hadn't objected because she was passionately in love with him and hadn't wanted to personally slur his character.

Hawthorne swiftly went on to a different tack:

"Did you love Clark?" he asked.

"Yes."

"Did you love Lancaster?"

"Yes, years ago."

"Do you love Lancaster now?"

"No, but I'm very fond of him," Jessie replied.

Hawthorne's voice thundered around the courtroom as he pounced: "Mrs. Miller, did you betray Captain Lancaster in your love for Clarke while Lancaster was in the west sending you all the money he could beg, borrow or steal?"

Carson leaped to his feet, objecting to the word "steal," and Judge Atkinson upheld the objection. Jessie answered that she admired Lancaster because he had defended her at all costs. Undaunted, Hawthorne kept attacking.

"Would you die for Captain Lancaster?"

"Yes."

"Would you lie for him?" Hawthorne barked.

Jessie paused for a beat and then shot back an emphatic: "No! Because you would know I was lying." Then she attempted to explain their situation further. "You don't understand our situation. I am sure that being in love and loving a person are quite different. I was not wildly infatuated with Captain Lancaster. We worked together and had mutual friends." When she was dismissed from the stand at the end of the morning's session, a distressed Jessie could be heard weeping to a friend at the door, "They're crucifying me."

But Hawthorne had a point to make and in the afternoon, sensing a weakness, he closed in on his prey as he attempted to prove how far Bill would go to have Jessie. Knowing everyone in the room was already aware Bill had written the suicide notes in an effort to protect her, he asked:

"Then if Captain Lancaster has committed forgery in the two suicide notes, does it increase your fondness for him and raise his code of honor in your eyes?"

"No," Jessie admitted, standing her ground fearlessly. "It was a foolish thing to do."

"Would Lancaster give you any money he had if you needed it?" Hawthorne persisted.

"Yes."

"Would he die for you?"

"Yes."

"Lie for you?"

"Bill doesn't lie."

"Steal for you?"

"He doesn't steal."

"Wouldn't he steal a chicken for you?"

"No!" Jessie snapped. "It was a duck!"

The courtroom collapsed in helpless laughter. It was an unfortunate image, even if it was true. The court had already heard Lancaster's diary entries about stealing chickens and rabbits to put food on the table, a petty crime in which probably more people than Lancaster in that room had indulged in those hard economic times, and so instead of scoring a point against Lancaster's reputation, Hawthorne lost his grip on both his witness and the room. He'd gone one question too far. In a desperate attempt to make up lost ground, he fell back on reading more diary excerpts in front of Jessie, but she didn't let him see her vulnerability again. "We went through a lot together," she commented about Lancaster. "We were pals. What I had was his. What he had was mine. We trusted each other. He was always ready to shield me in every way. I would defend and protect him at all costs. Bill did not thrill me but I was happy with him."[36] Hawthorne could see there was no more to be gained now and so he handed over to Carson for cross-examination. Carson let Jessie remind the court about Haden's tendency to lose his temper, Peggy Brown's tendency to always show up at the house bearing lots of liquor and Jessie's conscientiousness about regularly sending money home to Lancaster's wife and children. Then she left the stand and he recalled Special Officer Hudson for some cross-examination. Hudson grudgingly admitted to Carson that he had in fact known about the relationship with Peggy Brown, having found a telegram from her in Jessie's house. This was the telegram Jessie had found screwed up in a ball in the trash when she had returned to the house after her police interrogation. As the police had been the only people to enter the house while Jessie was being held in the cells, Carson promptly accused them of either deliberately or carelessly attempting to suppress a piece of evidence that was valuable to the defense's case. Everyone involved pleaded ignorance, but as Jessie could testify she had seen it sitting on the mantle the night of the tragedy, it was clear one of the investigators had read it and made a decision about it. Jessie would later estimate that it had been this revelation about possible police bias in the case that finally swung public opinion in Bill's favor.[37]

Rather than leave Jessie's tense drama with Hawthorne hanging in the air and in the minds of the jury at the end of the day, Carson decided to open for the defense in the afternoon by introducing Lancaster to the jury before the day was out. Some biographical facts could possibly distract them from Jessie's dramatic testimony and give them something else to listen to and think about. Bill's English-accented voice remained perpetually careful, calm and measured as James Carson took Bill through his life story to date, making sure that the jury knew he had been a respected member of the defense forces in two countries. Bill recounted how Jessie and he had met and of how their friendship had grown intimate during their flight between England and Australia and that after their arrival in Australia they were constantly in the public eye and so had little opportunity to see each other. Since then, he professed that his love for Jessie had increased during the years. He then went into detail about his financial and aerial bad luck up until his departure west with Tancrel and Russell, at which point court adjourned for the day, now aware of Lancaster the person rather than only Lancaster the villain.[38]

On Tuesday morning, Lancaster was back on the stand adding to that person with Carson's able assistance. Throughout his testimony, though, Lancaster persisted in using Jessie's married name, seemingly oblivious that in doing so he was continually reminding the court that he, a married man, loved a woman who until recently had also been married. Nevertheless, keen to develop Lancaster's character further for the jury, Carson led him through testimony that the relationship between him and Jessie had become intimate somewhere in Persia on the flight from England and that he knew he was in love with her by the time they landed in Australia. Their partnership had survived good times and bad and his love for her had grown, he continued. He testified about meeting Haden, about the arrangement between Haden and Jessie to write the book that had included Haden moving in with them, and that Haden and he had shared the same sleeping area. Lancaster then took the opening shots in the ensuing open season on Haden Clarke, the one person who couldn't defend himself, by claiming that he'd been in the house when Clarke had entertained other women while Jessie wasn't there. Nevertheless, before he'd departed to fly west, he had taken Clarke to one side and had warned him about Jessie's tendency to "weakness" when drinking, and had asked him to protect her. When Carson asked Lancaster to explain "weakness" he replied that when Jessie "drank heavily her conduct was not normal." There had been parties when everyone had far too much to drink, he declared, and the guests on occasion had included Peggy Brown, who'd told him that she expected to marry Clarke. Despite all this, Lancaster had left a lonely, neglected and desperate Jessie at home with a handsome, younger, smooth-talking Haden Clarke who, though, had promised Lancaster to look after her in such a way "as to make me remember him for life." Whatever might be said about Haden Clarke, he had definitely kept that particular promise.[39]

Lancaster went on to assert he hadn't believed Russell's claims until he'd received the actual letters from Haden and Jessie telling him they'd fallen in love, but rather than being angry, he protested, he'd simply been puzzled and unsure of what to do. He'd finally sent a telegram wishing them luck and asking them to postpone the wedding until he could be there as best man. Before he left St. Louis, he admitted, he'd bought a pistol to replace the one of Huston's he'd pawned and had merely loaded it in Nashville because Huston's weapon had been loaded when he'd given it to Lancaster. Carson then led Lancaster through an account of the failure of his business relationship with Tancrel and Russell as he became more suspicious of the nature of the business and more convinced he couldn't trust them, and he took him through receiving the letters from Jessie and Haden, his return and their reunion back in Miami.[40]

Keen to leave no doubt for the jury of his defendant's integrity, Carson later in the day called a series of character witnesses to the stand, once again to leave a certain impression in their minds overnight. Keith Bon, a Singapore rubber broker, stated he'd rushed all the way from Paris to testify that Lancaster was an upright man. Then a group of well-known pilots who had all flown with Lancaster and had traveled some distance to be there testified in favor of Lancaster's character and reputation. They included Congressional Medal of Honor holder Frank Upton, fellow fighter pilots Lieutenant Irving "Bing" Boyer and Joseph Ince, aerial circumnavigator Clyde Pangborn, and Rex Gilmartin, the commander of the Aviators Post of the American Legion in New York. To a man, they all praised Lancaster's integrity.

When Lancaster took the stand the morning of 10 August, the seventh day of the trial, he faced Carson's direct question:

"Did you kill Haden Clarke?"

Showing no surprise at all, Lancaster calmly answered: "No."

Carson wasn't surprised, either, of course, and quickly proceeded to take Lancaster, via a series of questions, through the emotional and tense events of the evening prior to the shooting, and he gave a clear, fluent and detailed account of not just events but also entire conversations, suggesting that he might have put all that time spent in custody to very good use indeed in committing his story to memory and rehearsing the narration. Lancaster professed he'd been reluctant to discuss with Hawthorne the final conversation he'd had with Haden and was still reluctant because he wanted to protect Jessie and Mrs. Clarke. Carson duly absolved him of responsibility and so

Ten. An Awful Thing Has Happened

Lancaster proceeded to tell the court that he'd actually been impressed by Haden's sincerity. Haden had confessed to having a sexually transmittable disease, to being still technically married and that he'd lied to Jessie about his age and education. However, he had expressed remorse and regret and had declared that, although he'd had his share of affairs, he was absolutely in love with Jessie and would do all in his power to make her happy. Obviously forgetting he'd heard Jessie testify that it was a few hours prior to this when he'd advised the couple to wait until Haden's situation was clarified, Lancaster now claimed it was during *this* conversation he'd first learned Haden was not yet legally divorced. After all that, Lancaster said, he was really tired and so they'd agreed to discuss everything with Jessie in the morning. Clearly, Lancaster wanted the jury to remember him as Jessie's honorable protector. One can imagine that he must not have been able to believe his luck on being handed such powerful information. The unsaid words "or I will" hung in the air at the end of his advice to tell Jessie. "I knew before I went to bed that Haden could never marry Chubbie, and I think he knew it too," Lancaster assured the jury, thus establishing for them the perfect reason why Haden would have committed suicide. As the only survivor from that room that night, Bill had no better opportunity to seize the chance to write his own victorious history from which he could emerge looking like a Sir Galahad, and in this case there was an additional reason: his life depended on it! Even Clarke's last words, according to Lancaster, were a compliment about Bill's honesty: "You're the whitest man I ever met."[41]

During cross-examination that afternoon, Hawthorne then took Lancaster through the events surrounding the shooting itself and its aftermath. In answer to his questions, Lancaster claimed he'd been awakened by a loud noise that he'd at first thought was a window banging. He denied firing the fatal shot, although he admitted to placing the loaded weapon on the table between their beds that night. After discovering the bleeding Clarke, Lancaster coolly admitted that before raising any alarm at all he had sat down and typed out fake suicide notes, which he had callously attempted to have the badly wounded man sign. To demonstrate to the jury the ease with which this could be done, Hawthorne sat Lancaster at a table in front of the typewriter that had been in the bedroom and had him type the notes while timing him. It took all of two and a half minutes to type both, refuting Lancaster's earlier claim it had taken five. Then, asked why he thought Clarke would have killed himself, Lancaster maliciously declared: "I think he killed himself with good intentions because of an illness he had contracted." Having been right there in the courtroom when Hawthorne queried Jessie on the implications of the word "illness," Lancaster had just deliberately damaged his own best friend's reputation in an effort to save his own. Then he uttered perhaps his most his breathtakingly fatuous statement of the entire trial: "I see his act as a beautiful thing." Unbelievably, the courtroom audience agreed with him so loudly, cheering and applauding, that Judge Atkinson sent the jury out and then warned the audience he'd clear the courtroom if there was another demonstration of that nature. But Bill had made his point, and after at least nine hours on the stand over three days he finally stepped down.[42]

Facing the courtroom the following day, Carson reminded the audience of yesterday's reaction and asked them not to hinder his client's chances for a fair trial. Then he proceeded to launch into his own attack on the character of the late Haden Clarke in order to demonstrate just how much of a lowlife Haden had been. In this way he could paint Jessie as a woman of weak character, easily led astray when left on her own, and Haden as an experienced seducer and con-man who'd taken advantage of her. According to this picture, it would be inconceivable for a man of honor and noble character such as Lancaster to descend to the level of murdering a person as worthless as Haden unless he was in fact performing a deed that was in itself honorable: saving the woman he loved. To this end, Carson introduced into evidence love letters to Haden signed by Virginia, Peggy, Eleanor and a few other women. Virginia had written that she was "planning my future with you. I dream about you every night, my wonderful lover. You must never leave me again." A local landlord was called who testified Haden had rented an apartment from him accompanied by a "wife" named Clifford and that he'd produced a certificate for this second alleged marriage.[43] Haden was pilloried as a morose young man with suicidal tendencies who was addicted to narcotics. Dick Lavender, who had occupied the same house with him in New Orleans in 1931, testified that he and Haden had discussed suicide back then and that Haden had declared that "if he ever got in a jam he would shoot himself in the head." He added that he'd known Haden to consume drink and drugs and that he was "afflicted by the same illness that he had before he was shot," that is to say, a sexually transmittable disease. A local restaurant proprietor, Mrs. Alma Troop, who'd often given Haden a free meal when he was broke, claimed that he'd been constantly depressed and once told her life would not be worth living without Jessie.[44] Richard Richardson, author of the play about Rasputin, testified he and Haden had once been discussing that the difficulties of murdering Rasputin could have been avoided had the conspirators shot and killed him, "the way I'll do it," Haden had declared, further clarifying that a doctor had told him "it was a mistake to shoot myself in the temple, as it could sever the optic nerve and cause blindness but not death. The right way is a little above and behind the right ear."[45] If these were Haden's friends, Carson certainly hadn't needed to find any enemies. Finally that day, Carson undermined Tancrel's testimony by producing Deputy United States Marshal J.P. Moe, who told the court that while Tancrel had been in Federal custody on a charge of impersonating a naval officer Moe had heard him threaten to kill Lancaster

Ten. An Awful Thing Has Happened

if they were placed in the same cell. "I'll do everything I can to see that Lancaster burns," Tancrel had declared.[46]

As if the ghost of Haden Clarke had not been dealt with severely enough, further embarrassment awaited him on Friday, 12 August, in the form of "Dr." Albert H. Hamilton, who was, according to Colin Evans in his book *A Question of Evidence*, the "undisputed master of evidence manipulation."[47] In a truly ghastly piece of courtroom drama that belonged more in the nineteenth century than the twentieth, Haden's bullet-pierced skull was held up in court by this forensic charlatan to the shock of the jury and the audience alike. Understandably, there were audible gasps from the audience and one or two fainted, yet judging by his silence on the matter, Judge Atkinson didn't seem to find this behavior in any way unusual. Fortunately, neither Jessie nor Mrs. Clarke was in the courtroom that morning. When Hamilton strode to the stand, this small, confident-sounding man gave the impression that he knew what he was talking about. He had been, he claimed, a criminologist who had appeared in 296 murder cases over some forty-seven years, but that was rather loosely applying the term. Hamilton had begun his career as a producer and seller of patent medicines who, as a shrewd student of human nature, soon figured out that a tidy profit could be realized from the public's growing fascination with and awe of things scientific, as well as the fifty dollars a day plus expenses for appearing in court as an "expert" witness. With credentials entirely of his own devising, he soon moved into the field of ballistics, having read some early reports on bullet identification. However, while there might be a sucker born every minute, as a number of con-men are supposed to have said, you can't fool all of them all of the time. Hamilton had found himself in serious hot water in his home state of New York back in 1915 when his erroneous testimony regarding the identification of bullets nearly led to the execution of Charles Stielow for the alleged murder of his neighbor and housekeeper. Six years later Hamilton was again in trouble, this time in Massachusetts, when he was caught attempting to switch gun barrels to shift the evidence in his favor while testifying for the defense during the famous Sacco-Vanzetti trial. Nevertheless, here he was again at the other end of the country, having offered his services to Carson to appear for the defense, attempting to prove Lancaster's innocence in his typically flamboyant style by pointing out that self-inflicted gun wounds do not always display external powder burns. A man of Carson's knowledge and experience in matters of law must have been aware of the shadows cast on Hamilton's reputation by now, shadows that could easily have damaged a defense case based on morals and honor, so having Hamilton testify in this case hints at a desperate gamble.

Hamilton did not let Carson down. Haden's actual skull had been introduced into evidence earlier when brought into the room inside a box by Dr. M.H. Tallman, one of the four doctors constituting the autopsy commission panel who had examined Haden's exhumed body. Tallman had merely proceeded to point out the entry and exit wounds and then respectfully re-placed the skull back in the box. At least one reporter watching Lancaster during the display noted that he seemed "visibly unaffected by the presentation of the skull" and actually "leaned forward with interest."[48] Tallman, along with Drs. Donald Gowe and Walter Jones, gave their opinion that the pistol had been pressed firmly against the head but were prevented from giving an opinion about whether Haden committed suicide by an objection from Hawthorne that they were not qualified to do so.

Unlike Dr. Tallman, Hamilton wasn't going to let respect get in the way of good courtroom theater. Dramatically hoisting Haden's yellowing skull from the box as if he were Hamlet about to soliloquize on Yorrick, Hamilton rolled it around in his hands while he expounded, at one point getting so carried away that he nearly dropped it, much to the horror of the room. The bullet had been fired across the head and slightly backward, he declared confidently, pointing out that the absence of surface scorching or powder burns indicated a sealed contact; that is, the gun had been held so tightly against the head that explosive gases had been prevented from escaping. Hamilton claimed tiny bits of lead had been found around the wound and that an explosion inside the skull had caused fractures and had left powder and gas marks on the inside tissue that had "ballooned" away from the entry wound. With evidence that only pointed to the gun being fired in close proximity to Haden's head, not to who might have fired it, Hamilton then delivered his own verdict: "This man took his own life. It is an absolute case of homicide!"

There was an audible gasp from the room as Carson jumped to his feet, crying: "Homicide?"

Caught completely on the wrong foot, Hamilton hesitated a minute, then laughed as he corrected himself: "Did I say homicide? No—suicide. I was thinking of something else."[49]

There was no doubt, he went on, that the cause of death was suicide because of the angle of the entry and exit wounds and the discoloration of the skull that indicated powder burns. Then, for further effect, he picked up the exhibit pistol and, holding it to his head, idly snapped the trigger as he talked. He revealed that Haden's brother, Dr. Beverly Clarke of New York City, had actually joined him in examining and testing flesh samples taken from the skull to determine if power residue was present. By the time it appeared as an evidence item in the court case, the .38 Smith & Wesson with its four-inch barrel had been cleaned inside and out, possibly by the clerk from the Miami Bureau of Identification who had test-fired it for ballistics purposes, although he fell back on the "I don't remember" defense when Carson queried him on the stand. However, Hamilton claimed that sufficient blood and tissue residue had remained on the end of the barrel for his examination and that some hairs had been present on the gun sight. After all that theater, the judge, jury, lawyers and defendant were probably glad to have a break outside the court in order to visit the actual upstairs murder scene in person. There the defense made sure it was pointed out to the group that the right side of Haden's bed was only a few inches from the wall; it would have been difficult for anyone to have come up on that side of the bed to shoot him.

On Saturday, the defense and the prosecution both still had some last loose ends to tidy up. Hawthorne called back Lancaster, for one final time, who vehemently denied he'd told Mrs. Clarke since the tragedy that he'd been so confused on the night of the shooting he sometimes thought he had killed her son. Then, Drs. Tallman, Gowe and Jones, who had comprised the Haden autopsy panel, each testified that the pistol had been firmly pressed to Haden's head. However, when Carson then asked them for an opinion on whether Clarke had committed suicide, Hawthorne objected, forcing each doctor to admit he was not qualified to give such an opinion but could only speak in terms of the results of their autopsy observations and what those results most probably indicated. Finally, at three in the afternoon, after testimony from twenty-four witnesses, the defense rested.

Then it was time for some rebuttal testimony. It was Hamilton's turn to be put under the microscope. Dade County investigator C.A. Peterson testified that he was familiar with Hamilton's reputation as a criminologist.

"Do you know of his reputation as an expert?" queried Hawthorne.

Carson objected on the ground that only people living in Hamilton's home area would be qualified to testify to his reputation but was overruled.

"Yes, I know his reputation," Peterson answered.

"Is it good or bad?" Hawthorn wanted to know.

"Bad," Peterson replied.

Hawthorne could never resist a parting shot: "In view of your last statement, would you believe him on oath?"

"I would not," Peterson emphatically declared.

With no further introduction needed, really, Hawthorne then invited Hamilton to take a very hot seat and grilled him unmercifully about his unethical involvement in the Sacco-Vanzetti case, winding up by reading into the record a long report about the matter.[50]

Rebuttal continued on Monday. Carson re-called Lancaster's wartime pilot friend Captain Joseph Ince, who said that he'd been in the El Paso Hotel when Lancaster had said he was glad Clarke was taking care of Jessie and denied Tancrel's claim that Lancaster had said he'd seen hundreds of dead men and one more wouldn't make any difference, referring to Clarke. Hawthorne then set about restoring a little of Haden's character by calling a friend, architect Vladimir Virrick, who testified that after Clarke had spoken to Lancaster on the telephone in his hearing Jessie had commented that Lancaster was returning and there would be trouble. Airport mechanic James Enrico contradicted Jessie's earlier testimony by declaring that Lancaster had been anything but cordial when he greeted Haden at the airport. Mrs. Clarke was too ill to testify as planned, and so Hawthorne had Dr. Donald Gowe return to the stand to deny Hamilton's testimony that human hair and brain tissue found on the gun barrel could be identified under a microscope, followed by Charles Morgan, a Miami chemist, who declared he'd examined the gun barrel on which Hamilton had claimed to have seen human hair. It was no such thing, he asserted: it was cotton fibers.

Finally, late in the morning of Monday, 15 August, the defense rested their case and closing summations began under conditions that were a lawyer's dream: there was no stipulated time limit. Henry M. Jones, the Assistant State Attorney, began the summation for the prosecution team with criticism of Lancaster, which may not have pleased Judge Atkinson, who made the unusual gesture of turning his back to the courtroom. "This man, whose woman was stolen by Clarke, planned the whole murder carefully beforehand," Jones began regardless, "but he made it look too good, just as his counsel has done with this case." Then he turned on Jessie, declaring, "From this woman, weak physically and weaker morally, came testimony tending to show that this awful tragedy was suicide. One of her lovers is dead and she is doing what you would expect. She is protecting the lover who is still alive." He claimed Lancaster's testimony that Clarke killed himself in an honor suicide was "too smooth, too perfect," a "self-serving profession of innocence" that he had carefully memorized like a schoolboy. Lancaster's motive for murder was "his planned revenge because Clarke had stolen his woman," Jones declared, and it was written all through his diary, showing in his mental agony when he learned of the engagement. This was a case being watched internationally, Jones concluded emphatically, and the eyes of the world were upon them.[51]

When James Carson rose to his feet after lunch that day to begin his closing argument, the only one from the defense team who would speak, the room knew this would be a much longer speech than the previous one, but at least it would be an erudite oration with characteristic southern flourishes, and they were not disappointed.[52] Carson began by comparing Lancaster to, of all people, the great Civil War leader General Robert E. Lee. Using a nineteenth-century speech by Senator John Daniel of Virginia, he pointed out that although Lee had been stripped of everything, he had still "towered above the earth and those around him in the pure sublimity and strength of that character." Lancaster, too, he declared, far from home and charged with murder, had been stripped of money and friends and confined to a cell, yet

Ten. An Awful Thing Has Happened

he'd stood above it all "in such pure sweetness, strength, unselfishness, and sheer nobility of character with the brilliance of a gem against the dirt, mud muck and filth which form the sordid background of this case." If indeed Lancaster was being charged with adultery, writing bad checks or stealing chickens, he reminded the court, they wouldn't be sitting in this room. They were there to judge the defendant specifically for murder beyond a reasonable doubt. His client, he reminded them, was not on trial for his love or he would be convicted, nor was he on trial for being one of "this sordid triangle" or he would be convicted. Rather, he was on trial for killing Haden Clarke, and because the State had not proved his guilt beyond reasonable doubt he should not be convicted of that.

Carson resumed his closing argument the next morning by reminding the court of the closing words of Henry Jones. However, Carson declared, managing to refer at the same time to both Napoléon's speech to his troops in Egypt and the Magna Carta, not just the eyes of the world were upon them but the eyes of ten centuries of history. He asked the jury that they not convict a man based on circumstantial evidence that could possibly fit any other theory or in which there were any gaps in the chain of circumstance. Then, using some case history examples from England as well as America, Carson illustrated for the court the possible ambiguities of such circumstantial evidence as they had been hearing and in what way they could lead to wrongful convictions. It was important, he told them, to approach this evidence from the point of view that Lancaster was innocent and had to be proved guilty, and he elaborated on the dangers of prejudice as to guilt or innocence. They were all sworn under oath to carry out a solemn duty, he reminded the jury, and being party to a wrongful conviction would be an act with which each of them would have to live. There had been some optimism within the room that the case might end that day, but by now it was five in the afternoon and so Atkinson recessed the weary court until the following morning.

Carson continued his summation by pointing out for the jury that the prosecution's case rested on five circumstances:

- Lancaster's "tremendous, unselfish and undying love" for Jessie,
- Lancaster's threats towards Haden testified to by Tancrel and Russell,
- Lancaster's purchase of a pistol,
- Lancaster's forgery of the suicide notes,
- Lancaster's queries as to whether Haden Clarke would speak again.

These five "disconnected and sketchy circumstances [were] in no way sufficient to give you any reason for believing Lancaster killed Clarke," Carson stated. The defense had accepted these circumstances and explained that they were but isolated events; they did not provide the whole scenario, and it was dangerous to take unrelated facts as proof. The State was stressing motive, he went on, but really they were saying that if Lancaster didn't kill Clarke then he ought to have killed him. However, what might be a motive within this culture might not be one within another. Just because someone from Kissimmee may avenge wrongs this way did not mean an honorable Englishman would.

Carson then reviewed the list of witnesses and their testimony. Hudson, the policeman, had denied he'd wrapped the gun in a handkerchief but claimed that he'd put the gun in its box and then put the box in his hip pocket. "You can look at the box," Carson directed the jury, "and if you gentleman will pardon me now, I will just put this brief case in mine [hip pocket]." A chuckle rippled through the audience. Hudson, he reminded them, had attempted to destroy or lose evidence that would have supported Lancaster and had falsely claimed to be the brother-in-law of Peggy Brown, who had spent a night or two at the house with Clarke, and Hudson as the only investigator on the stand did not have anyone else to corroborate his story, in which there were obvious mistakes. "If the time has come in this country," Carson ground out as he roundly condemned Hudson's actions, "when a dumb, thick-headed cop, without intelligence enough even to be foxy, with a motive that is undeniable, with the type of mind that always says guilty, can put you … in that chair, then it is time somebody did something about it."

Then Carson started in on Jessie's qualities as a witness, having reminded the court that during this trial they had heard "many sickening and unusual and sordid details concerning sex." While his comments about Hudson may have been justified, Carson's derogatory comments about Jessie that followed can only be justified in terms of their early discussion and his warnings to her that he would have to paint her as a scarlet woman, and Carson proceeded to splash on that color in thick, broad brushstrokes. It was well that she was not in the room to hear him declare that despite her bravery, courage and matchless gallantry, he was sorry for her because due to some pathological condition, she was one of those women who were "utterly unable to live up to the standards of virtue and chastity which you and I have been taught to believe constitutes the crowning virtue of the sex to which our mothers belong." This condition, this defect, had made it necessary for her secret weakness to be made public and exposed to the world. Nevertheless, he declared, she was telling the truth and had resisted the prosecution's attacks, in doing so demonstrating "the matchless and dauntless courage that allowed her to be the first woman to fly half-way around the world." As he sat there, listening to the moral reputation of the woman he supposedly loved being destroyed, Lancaster should have been ashamed of himself that this had been deemed necessary in order for him to escape conviction, but he didn't look very ashamed. Instead, the *New York Times* reporter thought it significant to note, "Lancaster was smiling as Carson discussed Mrs. Keith-Miller." When that journalist located Jessie waiting outside, she merely commented, "I am interested only in freeing old Bill. Nothing else matters."[53]

Carson then at last moved on to ridicule Dr. Deederer's demeanor on the stand and to accuse Hudson of wiping clean the gun that fingerprint expert Barker had testified had no prints or blood remaining on it by the time he received it. He pointed out that the prosecution had descended to using criminals as major witnesses: J.F. Russell was currently in the Dade County jail in the same building and had previously been in Leavenworth for people-smuggling, while Tancrel had been sentenced for falsely claiming to be a U.S. Navy officer. Worse still, Carson declared, the Dade County physician, Dr. Thomas, had claimed he'd carried out an autopsy that had revealed powder stains in Clarke's head wound and had written a report that had been presented to the grand jury, but Dr. Thomas had never been called to the stand because, Carson accused, either there had never been any initial autopsy at all or the report had been suppressed.

All in all, Carson summed up, while the State had started out to prove murder, they had ended by conclusively proving suicide. He asked the jury to compare the word of the jailbirds, "foul carrion birds, who come to us from the cells where they are charged with the betrayal of your country and mine," who were the State's witnesses, against that of the defense's "clean, brilliant and outstanding war-birds of the world, birds who come to us from the clean highways of the sky" to testify as to Lancaster's character. Carson then produced eight reasons for Haden's suicide that Jessie herself had written down: remorse, doubt Jessie would stay with him, financial worries, doubt of his writing abilities, no sex life, too much responsibility, illness, and mood swings. After a noon recess for lunch, Carson finally started to conclude by giving Haden Clarke some benefit of the doubt. He was a clever, intelligent young man, Carson, pointed out, who might have made a significant contribution if he could have found his way out of this situation. However, he couldn't, and then there was the matter of his disease. It would be "much kinder to the memory of the dead boy," Carson morally sermonized, "to believe that he consciously chose to commit suicide," that he'd "at last found somewhere within the recesses of his soul the courage to do the noblest act of a misspent life." Lancaster hadn't shot him; Clarke, who was broke, a chronic drinker, and a drug addict who had women problems and a recurring illness, was "an honor suicide—there was nothing else for him to do." So, he asked of those twelve good men, "Wasn't that enough to make any man commit suicide?"

Then Carson actually felt it necessary to launch into a defense of his own witness, Albert Hamilton, lauding him for testifying at his own expense and for being an expert in the fields of weapons, photography and witnessing.

Hinting at casting stones in greenhouses, Carson pointed out that his chief critic, Peterson, had himself made a few mistakes. Surely they were here to try Lancaster, he pontificated, not the worthy Hamilton, who had only the best interests of proving a man's innocence in mind. Finally, after pointing out that while Lancaster had made mistakes that didn't mean he was guilty of murder, he once again asked the jury to weigh the characters of the men on each side, the jailbirds of the State with the war birds of the defense, and find Lancaster not guilty and so "send word back to old England from whence we get our Common Law that in American courts justice is administered in the high, fair and solemn fashion that our ancestors won in that soil by their blood." Finally, after some six solid hours of summation over two days, Carson at last sat down.

Hawthorne then rose for his final words, reminding the jury that Lancaster was guilty "of violating several of the Ten Commandments ... and Mr. Carson says this man's moral code shines like a star on a background of filth."[54] Lancaster's diary, Hawthorne reminded them, contained a series of threats against Haden, who had, he reminded the jury, stolen Miller's love. Hawthorne concluded his plea for Lancaster's conviction by asking the jury to judge him guilty of a cold, calculating murder. "The defense tells you that this man [Clarke] committed suicide," he declared. "State witnesses, uncontradicted, have shown a clear case of murder, contemplated clear across a continent, by a man who is capable and cold and crafty. His diary shows his intent, day after day, to kill Haden Clarke." Then he rounded on Lancaster. "Do you think," he asked the jury, "that when Lancaster came back here and found Clarke and this woman in love he slunk away like he says he did and just planned to leave town? No, he didn't! The male of the species will fight to the death for the female he loves; the male animal of the field will do the same; so will the male bird; and gentlemen, so did the war bird—the war aviator sitting here before you."[55]

At 11:18 on the morning of 17 August, Judge Atkinson read his charge to the jury, reminding them that to find Lancaster guilty as a result of circumstantial evidence they had to find that the circumstances were inconsistent with innocence, and they left the room. So, members of the twenty-first century jury, let us also consider the main points of argument. Here we have a man who is desperately in love with a woman who, though a remarkably good friend if not his best friend, is by her own admission no longer in love with him. Despite everything he has been able to do and say, she's fallen for and is about to marry a younger, better-looking, smoother-talking and more morally conflicted rival for her affections. He's shady and dangerous compared to Bill's steady and stable, but Jessie has a long-standing preference for adventure and risk. On the other hand, Bill is also something of a gambler and adventurer, and he'd staked everything on this last hand in Miami. If

Ten. An Awful Thing Has Happened

Jessie left him, Bill stood to lose everything and would probably have to return as an ignominious failure to England, where he'd finally have to face his family, especially his father who'd been largely keeping Lancaster financially afloat. More than one person testified, for various possible reasons, that they'd heard Lancaster lose his temper about Jessie's relationship with Haden and in some way threaten him, and his diaries contained threatening entries. Lancaster himself admitted purchasing a pistol and ammunition on the way back to Miami and then loading the weapon and carrying it around with him into the house, after which he took it out and placed it on the bedside table within reach of a man who was clearly distressed and upset after a discussion about the unlikely success of his romantic intentions. Even if Lancaster did not pull the trigger himself, he certainly enabled the ensuing tragedy. In any case, given only Haden and Lancaster were in the room that night, any account of conversations or events was solely Lancaster's account; only the victor in this case could write the history of Haden's last words and actions for the jury and it would be an unusually honest man who would implicate himself under those conditions. Finally, Lancaster did admit he forged the suicide notes with some flimsy excuse that he'd done it to protect Jessie, but only when he knew the District Attorney's investigator was about to reach that conclusion. Lancaster never did explain why he was so concerned whether the wounded Haden might speak except, perhaps, that Haden might say how he came to be shot in the head. It is, after all, virtually impossible to hold and fire a heavy .38 revolver with a four-inch barrel while lying in Haden's position in bed and achieve the bullet track visible through his skull. Not only that, despite an alleged interest in suicide methods, both court testimony and his mother's words clearly indicated Haden was renowned for his aversion to weapons and, unlike Lancaster, had never owned or actually held or fired a weapon in his life. Now, while the jury is out, how would you vote? Guilty or not guilty?

Four hours and forty-eight minutes later, the jury filed back into the courtroom having had to take only one full ballot. Their first vote, a partial ballot, had been taken barely fifteen minutes after they'd retired, but several jurors were still confused about the judge's instructions regarding reasonable doubt of guilt and had to ask for further explanation before voting. When clerk of the Circuit Court E.B. Leatherman announced the jury's verdict, "Not guilty," there was pandemonium as the crowd of predominantly female spectators packed into the room broke out in screams and cheers. One lady in the front row swooned in her chair. "Fashionably gowned women wept and shrieked in delight," the *New York Times* journalist observed. Another more philosophical journalist queried such an emotional female reaction to the verdict, pointing out that women had not always been so protective or sympathetic to those on trial, such as those women who sat knitting at the foot of the guillotine during the French Revolution. "Did women view Captain Lancaster as an avenging hero," the writer wondered,

> who—so the state charged—took into his own hands the law of "one man for one woman"? Did they see in this love triangle a romance in which they could picture themselves as the heroines? Could their act be called the expression of motherly protective instinct aroused by the fact that the British captain was a stranger within our gates whose sense of justice had been outraged? Were they merely voicing public sympathy for Chubbie regarding her as, in the words of a movie subtitle writer, "one of Fate's poor playthings"?[56]

Lancaster just quietly smiled as the crowd surged forward to grasp his hand and then he stood up, the room falling silent as he moved in front of the jury box to thank them for being "very patient with my case. You have had to listen during the long trial of the case to many things. I want to give you my heart-felt thanks for exonerating me." Then he clicked his heels and bowed to them and stepped away, but he was prevented from resuming his seat by another surge of well-wishers shaking his hand and slapping him on the back, as bailiffs, deputy sheriffs and police vainly attempted crowd control. After the room was finally cleared, Lancaster posed for photos with the jury and judge and then, gathering the reporters around him, he read another statement:

> I am very glad to find that I am free to go out in the open air once more. I owe a great deal to two gentlemen, Mr. James Carson and Mr. J. H. Lathero. I owe heartfelt thanks to the many friends who have rallied to my support and to the people of Miami who have accorded me, a Britisher in their midst, sympathy, kindness and justice. I consider my trial was eminently fair and impartial in every way.

His first desire after all this was to find a job, he concluded. When asked if his plans included Jessie, Lancaster was evasive. "Please do not ask me to answer that now," he begged off. "I don't know what my answer would be."[57] It was a far cry from the declarations of undying love that had been a feature of his testimony in court. Jessie had slipped quietly out of the court before the verdict was announced, along with Mrs. Clarke, hearing it from an adjoining office before leaving for her hotel. To the one reporter who saw her and asked her opinion of the verdict, Jessie said simply that she was delighted and had known "old Bill would come through."[58] Back in England, Lancaster's overjoyed parents quickly claimed that the spirit of Haden Clarke had spoken to them through a medium during the trial and reassured them that Bill was innocent.[59] According to the *Daily Standard* newspaper back in Australia, Jessie asked them to print a remarkably intimate message of appreciation to her mother:

> A mother's love and sympathy through all my ordeal helped me to keep me going. She was true to the name of mother. I wish tonight, with the trial and travail over,

Ten. An Awful Thing Has Happened

that I could put my head in her lap and feel her soothing hands caress my face, putting behind me all the cares and troubles as when I was a baby.[60]

Almost two years later, Jessie would claim to have been party to a letter that had arrived at the courthouse a few days after the trial had ended, addressed to James Carson from New Zealand and dated 24 August. The letter had apparently been written by a woman who said she had at one time been Haden Clarke's "secret girlfriend" because of her engagement to someone else, and in it she alleged that Clarke had been a dangerous man, a drug user who became very depressed to the point of being not only suicidal but also homicidal once the drugs began to wear off. "He said he would kill us both if I would not marry him," she declared melodramatically, "so she is lucky he did not kill her too." Consequently, although she had loved him, this woman had deemed it unsafe to continue their relationship, she concluded, and had since married her fiancé. Jessie claimed Carson said that if the letter had been received in time the trial would never have happened.[61] Although there may have been an actual letter, to give Jessie the benefit of the doubt, it is such a convenient validation after the event of her actions that it is almost too good to be true. The existence of such a document would immediately beg two questions: Why did Jessie wait to release the letter until so much time had passed that it no longer had any impact, and why does it not appear in James Carson's published account of the trial, written some time after he would have seen the letter and been aware of its significance? Jessie knew Haden for some time, sober and under the influence, yet she never mentioned Haden was physically violent with her or that she felt threatened by him in any way. Like most people, Haden had bouts of depression, mood swings, and occasionally lost his temper, but Jessie consistently described him as being a kind, considerate, intelligent pacifist. Then there's the timing of it. This story was released while Jessie was attempting to find a backer for her entry into the 1934 Centenary Air Race to Australia. Perhaps, once again, Jessie needed publicity.

As an aviatrix, Jessie Miller appeared at a time when attitudes of and towards women were changing rapidly, when there was, in the words of Prudence Black, a "complex articulation of discourses, as cosmopolitan image had to be played off against technical necessities ... and changing social norms."[62] Perhaps there was a price to be paid, because for a woman to become a heroine

> there's a transition to be made from the interior life to the exterior, into adventure, moving from a personal satisfaction to external achievement—literally a step from the feminine world into the masculine world. But when she became a public figure, she became an object of gaze, and that necessarily meant she had to be feminized again. Once she'd taken that step and become a public myth, she herself was turned into a woman again.[63]

During the Lancaster trial, the court was never allowed to forget that Lancaster had been a war hero and aviator, although he probably never fired a gun in anger and his later aviation career had been completely overshadowed by Jessie, yet they were never reminded that Jessie was a pioneer aviatrix who had a significant solo aviation career in her own right, who had won races and who had broken a number of records. Instead, they were presented with a woman, an Eve who had succumbed to a serpent and consequently destroyed a paradise. As she was a public figure, an object of gaze, it was important that Jessie be feminized again in the eyes of that public, even if it meant she would be depicted as a fallen woman, as a tragic woman, yet all Jessie had wanted was to be her own woman.

Eleven

England, My England

After the trial, Jessie was driven from the hotel to James Carson's house for her first reunion with Lancaster since the trial had begun. Then she left with Lathero to stay at his family home a few miles outside of Miami. Lancaster saw her only once and made a single brief statement to the press that he was "eternally grateful for her dauntless courage on the witness stand on my behalf."[1] Despite their mutual wishes to find work again, however, neither Jessie nor Bill would work in the U.S. again. Their futures were about to be determined by forces over which they had no control, namely, the U.S. government. By now, the United States Department of Labor had evidently concluded that, innocent or guilty, Jessie and Bill were no longer wanted in the country. Jessie had only been at liberty during the trial period courtesy of a $1,000 bond posted by Carson, and her postponed hearing regarding the illegal entry charges was quickly put back on the agenda for Monday, 22 August. As soon as Lancaster was acquitted, there was a brief hearing in Judge Atkinson's chambers after which his diaries were turned over to the Department. They subsequently refused to return the diaries and they've never been seen again. Jessie's illegal entry case was swiftly dealt with in a closed hearing on that Monday before Labor Department agent James Forrester, after which his report was sent to Washington for a final decision. That very same week, a grounded Jessie would have read that her former flying colleagues Louise Thaden and Frances Harell (now Frances Harrell Marsalis) set a new endurance record over Long Island of 196 hours 5 minutes, smashing Bobbi Trout's old record by 73 hours, and the Cord Air Derby from California to Cleveland, along with a number of other races, was in progress as a lead-up to the National Air Races there.[2] It must have all seemed so far away from Jessie now.

In true government fashion, September arrived while public servants still debated about whether to just deport Jessie and Bill or to indict them on charges of conspiracy to violate immigration laws, in which case they would be imprisoned and then deported. William N. Doak, Secretary of State for

Labor, expressed his opinion on 2 September: "We will decide soon whether we will deport them or indict them," strong words about Jessie in particular, who had merely been a catalyst for events that had then happened around her.³ They were accused of conspiring to smuggle Chinese over the Mexican border into the U.S. by air for a large profit of which Jessie was to receive a one-sixth share, which was palpable nonsense but gave the Department a reason to seek Jessie's deportation. In Lancaster's case, the Department was waiting for a medical report to make a final decision. However, probably because in all likelihood they wanted to avoid another public trial involving rather flimsy evidence and a man who'd just been heroically vindicated, they left the exit door ajar for a short time and gave the pair a chance to leave the country voluntarily.

For some reason, perhaps to make his own statement about Lancaster's character, James Forrester curiously announced late in September that during Lancaster's hearing, at which Jessie was not present, Lancaster had declared he'd marry Jessie "whether she wants to do it or not."⁴ By then, Jessie and Bill were reported to be in New York with Bill's mother, where Jessie had returned in order to sell her apartment and pack her belongings for shipment to England. When Lancaster subsequently met with the British ambassador in Washington, D.C., to ask him to intercede so that he and Jessie could remain in the U.S. until Lancaster obtained a Nevada divorce in order to marry her, the whole matter had evidently become a diplomatic potato that was much too hot to handle and the ambassador declined to hold it.⁵

Obviously this standoff couldn't go on indefinitely. While on the one hand, the U.S. government obviously didn't want Jessie and Bill in the country, neither could they conclusively prove the pair had knowingly sought to profit from a people-smuggling venture or that Jessie was in the country illegally. So, they switched charges to the more general one of "moral turpitude," based on the personal revelations during Bill's trial.⁶ By now, Jessie and Bill had pretty much run out of influential friends and funds and so the government reminded them of that exit door that had been left ajar. The pair took the hint and, on 7 October, the Department of Labor announced that the deportation warrants had been dismissed with the understanding Jessie and Bill would leave within a week, although Jessie would still be officially escorted out of the country under the cloud of that moral turpitude charge. Despite some brief speculation that Jessie at least might return to Australia while Bill reconciled with his wife, Jessie and Bill booked passage to England, although on different ships.⁷

So, on Friday, 14 October, Jessie arrived at the pier in New York City, dressed in a black-and-white-check suit and black hat with a spray of gardenias at her shoulder and smoking one cigarette after another. "Do you think I will ever live this thing down?" she wondered to reporters. "Will people and the press forget it? Can I do anything, fly the Atlantic or something, so that they will put another tag on me?" Asked about the statement that Lancaster would marry her, Jessie replied that it was "'improper comment' as he was still married to his English wife who refused to divorce him." She was then escorted on board the liner *American Banker* by a female immigration officer. Tearfully standing at the rail, she waved good-bye to her friends on the pier below, including Lancaster. "Bill," she called down, "make sure you wear an overcoat or you'll catch cold."[8] Assistant Immigration Commissioner Byron H. Hull quickly denied that Jessie was being deported, maintaining that the immigration officer was there merely to certify her departure.[9] A few hours later, Lancaster, too, was accompanied by an immigration official onto a ship, the Cunard liner *Scythia*, for the long voyage back to England.

In contrast to her departure from America, Jessie's arrival in England was shrouded in secrecy. Her name did not appear on any passenger manifest and she kept to her cabin throughout the voyage, refusing to talk to anyone. Other passengers on the *American Banker* were completely unaware of who she was. Bill arrived in Liverpool on 24 October and raced to London to help Jessie elude the crowds that were gathering at the King George V Dock to catch a glimpse of her. As the ship was negotiating the lock about an hour prior to docking, Jessie disembarked over the side by a ladder and was driven away by Bill, who'd been waiting there with a car. It was winter and the middle of the Depression, and England was a chilly and lonely place. Jessie went to live with her aunt in Hampstead and nearly died of the cold. Bill returned to live with his parents, claiming to be planning another long-distance flight and working on a book.[10]

Jessie did not stay silent for long, however. Perhaps she was tempted by the money, which she would have needed, or perhaps she really did want to have her say, but only ten days after she set foot on English soil the *Daily Express* began publishing a series of articles written by her about the events of the last few months. She'd chosen them, she declared, knowing the paper to be "fearless and an unrivalled platform." This was at last her chance for people to hear her voice, and she took full advantage of the opportunity. "I cannot describe the blessed relief of it," she wrote, revealing the intense pressure she'd been under,

> to be away from that oppressive atmosphere of suspicion, intrigue, the sensational lies of the "tabloid" reporters, the badgering, the unsought notoriety.... I no longer want to hide in doorways or look over my shoulder for the lurking "shadowers." I do not feel that every movement, no matter how innocent, is exaggerated for the sake of further persecution.[11]

She felt that in England there was "a sense of justice and fair play," which would allow her an honest hearing, and so now she would tell the true facts behind what had happened on the other side of the Atlantic where "no celluloid law court has ever surpassed the dreadful theatrical atmosphere of that Miami court, from the shouting and pantomimic speeches of the prosecuting attorney to the final thunderous applause when the verdict was announced." Mind you, that verdict had been in Lancaster's favor and that applause was an indication of the room's approval, and so Jessie really had received the reward for which she had bargained her good name. However, any concession to that would not have harmonized with her speaking her own mind about the country that had turned on her, coincidentally at the very time she'd actually been applying for U.S. citizenship. An airwoman's life "is necessarily unconventional," she wrote. "It takes her to strange places, brings her in touch with all sorts of people with whom ordinarily she would have no contact at all." She'd been an innocent woman in a strange land, whose knowledge of American law had been gleaned from movies, when she'd been thrown into a nine-by-four cell and then interrogated without access to legal counsel. When she was finally released, it was only to find her answers had been fed to the media and she'd already been tried by the public. This, she declared, was what the American justice system called "protecting the innocent."

Given a pulpit at last from which to defend herself, Jessie seized the opportunity and didn't let truth stand in the way. Although her defense counsel had warned she'd become the "scarlet woman" during the trial, Jessie declared, she had not wanted to abandon her best friend and had stayed to defend him, knowing to do so would also mean the end of her American aviation career.[12] In reality, she couldn't leave the country because she was still under government investigation, and she'd been subpoenaed to appear as a witness in the trial, so staying was not an option, although she could have opted not to support Bill. She hadn't really been in love with Haden Clarke, she claimed, which was not quite what she said during the trial; it was just infatuation induced by the strain she was under at the time and, besides, Haden was probably drugging her drinks when she wasn't looking and when he asked her to marry him, "I lost my head."[13] In the end, though, it had been Haden who'd lost his head that was turned into a courtroom exhibit; Jessie was still alive, although a little scorched from the martyr's fire. Her loss of reputation and career had all been worth it, she cried, because an innocent man had been vindicated. "I have made mistakes," she poignantly concluded, "but I have paid for them.... The rest remains in the lap of the gods."[14]

Well, the gods didn't seem to be very interested in what was in their lap right then. Despite the tragic death of her former co-writer, Jessie was still determined to write a book about her exploits and was eventually introduced to a Mr. Walker, a journalist with the *Herald* in London who agreed to be her "ghost" writer. For a while, she would write the manuscript out in longhand and then travel from Hampstead into London to give the material to him, but that method lacked immediacy and Walker said he'd have to interview her. To make that easier, Jessie rented a beautifully furnished bed-sit apartment in Oxford Terrace, off the Edgeware Road. However, that writing arrangement, too, evidently didn't work out, because, for whatever reason, no book appears to have eventuated. Still, Jessie was now living nearer Lancaster, who would drop by periodically to visit, and sometimes they'd go out to dinner together to talk about old times. They remained good friends, but there was no romance. "I didn't have the same feelings for him at all, but I would have done anything to help him," Jessie recalled. "There was no question of partnering up then, though. He would still have liked to, but I was through. There was no future for us. I was very fond of him but the romance had gone."[15] Some people, however, were apparently still not convinced. One night, Lancaster arrived at her door to take her out to a dance, for which occasion they were dressed up in evening clothes. When he rang the bell, Jessie grabbed her fur wrap and stepped outside into the winter-dark night. Suddenly, she reeled back from a stinging slap to the face delivered by someone who stepped quickly out of the shadows and then ran away just as fast. Lancaster chased after them while an angry Jessie went back inside, where, on checking in a mirror, she could see a wide, red mark across her face. A few minutes later he returned, out of breath, to apologetically report that the assailant had been Kiki, to whom he'd said some harsh words for behaving in that manner.[16]

Economic times were still hard, and jobs were as hard to find for pilots in England as they had been in the U.S. For a while, Lancaster thought he'd found a job with Hillman's Airline, but unlike Jessie, he'd never qualified for a commercial license, and so in the end they turned him down. Growing more despondent as time went on, he became convinced that the only way out of his troubles was to once more do something spectacular in an airplane. In July 1932, Amy Johnson Mollison, who in 1930 as Amy Johnson had been the first woman to fly solo from England to Australia, had set a solo flight record in a Puss Moth from London to Cape Town, South Africa, of 4 days 6 hours 54 minutes. The record was still in the public mind because Lady Mary Bailey, the first woman to attempt the flight, had taken off from Croydon in England in January 1933, in an effort to beat Amy's record only to go missing for four days until found by French patrols 300 miles east of the Niger River and 1,450 miles from Oran. Lancaster decided that breaking that particular record would reinstate his flying career, so he managed to talk his long-suffering father into financing him for yet another plane, and it was in this way that Lancaster once again became connected to Australia. The aircraft they eventually purchased for £700, another Avro Avian, was none other than the *Southern Cross Minor*, the plane originally flown by Charles Kingsford Smith in September and October of 1931 in a failed attempt to beat Jim

Mollison's nine-day Australia to England flight record. Kingsford Smith had to abandon his plan of flying the plane back to Australia because of medical advice and so he had put the plane up for sale. Although it had an adequate range of 1,600 miles for the trip, its cruising speed of 95 miles an hour was 20 mph slower than Amy's plane. To beat her record would take accurate flying and a punishingly tight schedule with little time for rest.[17]

Like any other long-distance flight, especially one over un-mapped North African desert, there were major risks. Once he crossed the coast of North Africa, he'd be flying south over the Sahara Desert, mostly at night. At that time, there was only one trans-Saharan motor vehicle track running north–south and Bill would only be able to see it in daylight distinctly enough to follow it from the air as a guide. Knowing he had very little night flying experience and that he'd have no navigation instrumentation in the plane other than a compass, Jessie advised Lancaster more than once that he should only cross the desert in daylight in order to keep his bearings while following that barely discernible track south. Bill would be flying over a landscape with no major topographical landmarks or differentiation, and so he'd have to know exactly where he was all the time.[18] Jessie felt deep in her heart that this was a plan put together too quickly by a man too desperate to "put himself right with the world," but there was little she could do.[19] This time she stayed out of the planning altogether. Conscious she was edging into her early thirties, she now wanted to hold on to that independence for which she had fought so hard and to be her own person. As far as she was concerned, this was purely a Lancaster affair.[20]

Lancaster, in his usual fashion, was cavalier about it all, although constantly embarrassed by his father, who insisted on supervising all the expenditure himself, haggling with suppliers and always trying to get cheap deals. Jessie remembered that in photos of him before he left Bill looked totally exhausted, "absolutely haggard. I saw him standing there with hollow eyes and cheeks. He looked like a death head. He looked exhausted before he started and I remember thinking, 'He's not going to make it.' "[21] As she hadn't been involved, Jessie didn't think she should say farewell to Lancaster at the airfield, and so she stayed away. Instead, he came to see her one evening just before he left, unable to stay long because of other appointments. He showed Jessie the photos of her that he would take with him in his wallet and asked if she would give him her watch as a memento, but Jessie pointed out that it was so small and delicate he'd never be able to see the face while flying the plane. He'd be better off taking a larger one with a luminous dial, she advised him. Bill reluctantly agreed and, waving good-bye, stepped out into the night. That was the last time Jessie saw him.

At 5:35 on the morning of 11 April 1933, Bill Lancaster took off from Lympne in his blue single-seat Avro Avian biplane. His mother was there, as always, along with a journalist or two and some aerodrome staff. It was a far cry from the large crowd that had turned out to see him and Jessie fly to Australia.[22] While his mother had waited up with him as he caught some sleep during the night, she wrote down some of her thoughts. As many mothers feel about their sons, for her Bill had never really grown up; he was still "Little Billee" and she reminded him in this letter, which he would never read, that as a child he had dreamed someone would give him a small plane in which he'd fly to Sydney to drop in on his brother John. His dream had eventually come true, although not quite in the way either he or his mother had expected. This time, though, she was so anxious about this flight she was "losing grip on my faith," and so she was taking strength from a spirit message for Bill that her late mother had sent to her through a medium: "Look not back upon the past with its deep shadows, but look forward into the future with smiles and trust." As she'd done with the *Red Rose*, Mrs. Lancaster had painted on the plane a silver cross with flowers; this time, though, they were forget-me-nots instead of roses. With that, his mother placed Bill in the hands of the angels.[23]

Although Jessie didn't go to the airfield, she nevertheless followed Lancaster's progress closely. This time, she knew as much about Bill's plane as he did. Unlike the Avro in which they'd flown to Australia, this aircraft was fitted with long-range tanks containing enough fuel for eighteen hours' flying, enabling Lancaster to fly across France, over the Pyrenees and the Mediterranean to reach Oran in Algeria some 1,100 miles away in one leg. His next stop would have to be at Reggane, an Algerian trans-Saharan refueling stop and he would have to fly that leg in darkness to stay on schedule. From there he would fly to Gao, on the southern side of the Sahara, and then cross Nigeria and the Bight of Biafra direct to Cape Lopez. After that, he would follow the African coastline for 2,500 miles to Cape Town. Of the 6,600 miles he'd be flying, 1,450 miles of that would be over desert using only daylight visual navigation. There wasn't even a light on the instrument panel; he just had a box of matches. His only other survival supplies were two gallons of water, a few chicken sandwiches, some beef extract and a chocolate bar. He had a pith sun helmet, a pair of yellow goggles, a silver horseshoe for luck, and in his pocket was his wallet containing twenty-two £1 notes, two ten-shilling notes and two snapshots of Jessie. In order to break the record, Lancaster had put together a very tight schedule that left no room for the vagaries of weather or airplane mechanics. "Madness! The flight was madness!" a journalist would write many years later. "Here was Bill Lancaster going to fly across the Sahara at night with only a box of matches to light up his instruments. In those days, all long-distance flights were touched with madness as men—and women— risked their lives in flimsy boxes of wire and fabric, trail blazers in a bid that mankind should conquer the air and outpace the birds."[24]

Despite being warned that he faced a 20-mile-an-hour headwind all the way to Oran that would have a serious negative effect on both his speed and his record chances, Lancaster took off anyway.[25] He quickly fell behind. When Jessie heard that he was so exhausted on landing in Barcelona he'd little choice but to spend the night there, "My heart sank," she recalled. "I thought 'that is the finish, he'll never make it now.'"[26] For him to be so tired so soon in the flight was not a good sign he would beat the clock. He was still behind time when he reached Oran in Algeria at 9:00 pm, over fifteen hours after leaving Lympne. Amy Mollison had done it in eleven. Due to refueling and insurance difficulties, Lancaster didn't leave Oran until 3:00 am 12 April, striking matches to check his position in the darkness until in the dawn light he intersected the curve of the trans-Saharan track and was able to land at Adrar. From here, he took off intending to overfly Reggane and follow the track the 800 miles due south to Gao, but by now he was flying into a sandstorm that blew him off-course. He had little choice but to return to Reggane to refix his position and refuel. There officials noted that he was clearly dangerously fatigued and begged him to rest, but Lancaster could only see himself falling farther behind and rested for only three hours, then waited to see if the sandstorm blew itself out. It was now around six in the evening, and officials warned Lancaster that if he lost sight of the track in the gathering darkness he'd have to fly a direct compass course over 500 miles of featureless desert to cut the track near Gao but he still took off, expected at Gao around two-thirty in the morning of 13 April. If he didn't arrive, the French said, they would send a convoy along the track to look for him. Then they stood and watched as Lancaster in his tiny Avian took off into the Saharan sunset and disappeared into the vast intractable wastes of the desert, never to be seen alive again.[27]

When Lancaster didn't reach his destination the French, true to their word, quickly instigated a search along the track, knowing full well that there would be only a limited window of time to find him alive in the scorching desert conditions, but neither Lancaster nor his plane could be located. In a period before GPS, emergency beacons or extensive radio coverage, if he couldn't be seen along his projected compass bearing no one actually knew where he was. Although they didn't know it at the time, he'd drifted away from his course bearing during the night and he was nowhere near it. Jessie knew that Lancaster's two-gallon water supply would only keep him alive for a few days at the most out there exposed to the desert heat, if he carefully rationed it. Only a miracle was going to save Lancaster, but none was forthcoming. Jessie tried desperately to immediately organize an aerial search from England, but no one was interested in funding it, and she certainly didn't have the money to personally fit out a plane to fly that far. Bill simply wasn't an important public figure anymore and, besides, there was that lingering shadow of suspicion from the trial. In any case, as Jessie knew, there was only one aircraft in the whole of England capable of flying that far anywhere close to her time frame: a single-engine Fokker FVII known as the *Spider* that belonged to the Duchess of Bedford, Dame Mary Russell. Not your average aristocrat, the Duchess had made her first flight in 1926 at the age of sixty-one, and only three years later she'd made a record return flight between England and India in eight days with her pilot, Charles Barnard. Then, in 1930, she'd made another record-breaking flight with Barnard from England to Cape Town, so he was familiar with the route. Jessie and Barnard knew each other well personally and he wanted to help, but he was firmly contracted for other flights already and, in any case, it would take four days to prepare the Fokker for a major long-distance flight over such a remote and dangerous area. Factor in the flying time on top of that and it was all going to be too late.[28]

So, stranded on a far shore from where she was unable to help her friend, Jessie could only agonize about his fate. "I had a dreadful dream last night that he was lying in the desert and crying for food and water," she told reporters. "I live with hopes that he is safe with natives in some village. I cannot bear to think that he is out there alone." Nevertheless, she would continue to try to find a plane and financial support with which to carry out her own personal search, she declared.[29] Incredibly, despite Jessie's efforts for his son, Bill's father was just as unpleasant to her as he'd always been. "Mrs. Keith Miller's efforts are very much against all our wishes," he pontificated. "Everything possible is being done for my son."[30] No doubt her worries and the concerns of Bill's parents and family weren't alleviated, either, as the search for him continued using planes and vehicles and searchers on the ground, by various rumored sightings of Bill living with natives at some Saharan oasis or village that persisted through October and into January.[31] Stories circulated that a white airman who was not French and answered to Lancaster's description had survived an aircraft crash in the Senegal region in the spring of 1933 and was still living with native tribesman. But nothing came of them. Lancaster might just as well have blown away on the desert wind; no sign was ever found of him and eventually, reluctantly, the search was given up and he was considered lost. Lancaster's parents claimed to have received spirit messages from their son "from the other side through psychic channels." Mr. Lancaster was convinced that his son had been caught in a storm and blown far off his course into Northern Nigeria, possibly in the territory of Katsina, because he'd had a horoscope made before Bill's departure that had shown bad weather on 13 April. He was so convinced of this theory that when the chief of that area, the Emir of Katsina, visited London in July of 1933, Lancaster sought an audience with him. The Emir, who was interested in aviation, actually heard him out and promised he would order a search along the borders of his land for any sign of Bill, although neither man really held any hope he would still be alive. Mr. Lancaster declared that Bill "did not suffer; that is a great relief to his mother and to me. Sooner or later we shall know everything. He and his machine will be discovered ... after an interval of nearly four months."[32] As time would eventually reveal, he could not have been more wrong—about everything. Jessie, on the other hand, had a different take on it. "Captain Lancaster died as he wished to die—flying. It was not aviation that killed him but the false accusation of murder," she commented, keeping that torch of injustice burning. "He started on the flight to Cape Town in an endeavor to rehabilitate his reputation, though his machine was unsuitable, and the chances were 100 to 1 against success."[33] Maybe she was right, but Jessie was in denial about the other possible cause of Bill's death, one with which she was so very well acquainted: Bill Lancaster's carelessness, which had nearly killed them both more than once. Instead, she theorized that Bill had probably been trying one last time to make enough money to buy that divorce so that he could at last propose to her, even though her answer would have been no. Always the eternal optimist, he'd never stopped talking about them being married, but Jessie had been through too much and simply didn't feel the same way any longer. Instead, Lancaster had in the end solved his own dilemma when he vanished into the empty sky of which he was so fond. As London club owner Trevor Allen wrote about seeing Jessie one night, "In that little woman I saw all the tragedy of the Great Adventure which ends in–Silence."[34]

Being Bill, though, he left one final problem for everyone else to sort out. Three days before he left, he'd visited a solicitor and made out a will. When it was opened towards the end of November, his family discovered to their horror that he'd made Jessie his sole heir. Mind you, the family fortune came to the grand total of £170 from his small life insurance policy, but for his mother it was the principle of the thing! His parents flatly denied ever hearing about the will beforehand, and they contested it in court, claiming the money should go to his children. "Bill's mother was so violent about it all," Jessie still recalled many years later, "and screamed the place down and everything about his starving children and so on and said I was a wicked adventuress.... she was so violent. She was quite mad." Overwhelmed by such an extreme reaction, Jessie decided any argument just wasn't worth it and let the Court award Mrs. Lancaster the money.[35] Not content with that, and evidently considering she had a reputation to defend, Kiki then took the London *Daily Telegraph* to court in a libel case for writing in an article that not only had Bill bequeathed all his property away from his wife, but also he had declared before he left America that he'd divorce her. Kiki evidently considered that readers would assume that meant an English divorce rather than an American one, for which the grounds were quite different and which she never intended to grant, and thus that her delicate English reputation had been questioned by such a suggestion. The newspaper humbly begged her forgiveness and settled out of court. Mrs. Lancaster graciously accepted their surrender and said she "was satisfied it was an honest mistake."[36]

On 23 May 1934, New Zealander Jean Batten became the fifth woman to arrive in Australia by air from England when she landed in her Gipsy Moth at Darwin at 3:30 in the afternoon in a time of 14 days 23 hours 30 minutes, ahead of Amy Johnson's 19 days 16 hours.[37] Her two previous attempts, earlier that year and in the previous year, had ended in disastrous accidents, but much like Jessie and Bill, she'd persisted. Soon after, Jessie revealed that she'd been searching without success for a year to find a sponsor for her own solo flight to Australia and was still trying, but her friends were poor. Ideally, she commented, she'd now like to combine a record attempt with entering the Centenary Air Race from London to Melbourne.[38] This race had begun as the brainchild of the Lord Mayor of Melbourne, Harold Smith, to commemorate the centenary of the Australian state of Victoria, after which it was taken over as an advertising billboard by its sponsor, chocolate manufacturer Sir MacPherson Robertson, in order to test the feasibility of a scheduled three-day air service between the two countries but also to compare the capabilities of the new metal-skinned aircraft for long-distance flying. Prizes would be awarded for winners in both speed and handicap categories: first prize in the speed section was £10,000 and a magnificent 18-carat gold two-handed gold cup, with £2,000 for the handicap winner. By July, Jessie was still the only Australian woman in the race, as the pilot of an Airspeed Courier entered by R. K. Dundas, who had been Sir Alan Cobham's agent in India, along with co-pilot Mrs. Neil Fergusson. The two women and their airplane progressed far enough through the entry process to actually be given a race number, 36, and were among the final list of sixty-four entrants from fourteen nations within which were few other airwomen, only English entrant Amy Mollison (nee Johnson), who was flying with her husband, Jim, in their Comet and American entrants Ruth Nichols in a Lockheed Altair, Louise Thaden in a Beech, Laura Ingalls in a Lockheed Orion and Jacqueline Cochran in a Granville R-6H. An article in *The Sydney Morning Herald* celebrating this group of women did inevitably touch on what they could be wearing. "The competitors will pay little attention to clothes," the journalist observed, noting that lack of time to bother about it and lack of room for luggage would be the main factors and reminding readers that Jessie had arrived in Australia in a white shirt, shorts, flying coat and helmet. New Zealander Jean Batten had arrived in just a white flying suit, carrying just a green comb and a cake of soap. "It is thus apparent," the journalist summed up, "that women who embark on long-distance flights have to sacrifice very much indeed." Nevertheless, they added, many had said that flying "was a game for men alone, but the women of almost every country have proven that their talent is well suited to flying."[39] By the beginning of September, when the start date for the race was announced as 20 October, it was already being hailed as the "greatest air race the world has known."[40]

By early August, Jessie had changed to a Lockheed Vega with American airman James Woods as co-pilot, but by the end of the month a number of entrants, Jessie among them, were having financial problems. Despite public appeals for funding assistance in newspapers, she still couldn't raise the necessary funds. She even tried a personal plea to Sir MacPherson Robertson himself, pointing out she was the only female Australian entrant and offering to insure her life with Robertson as beneficiary as collateral for assistance to cover the final £1,500 for insurance, oil and fuel, but Robertson's reply was a blunt "Unable to help." Justifiably upset that even with her aviation record she was still unable to find a sponsor, Jessie complained to reporters that "I have approached advertisers and financiers, offering them any conditions, but it is apparently hopeless. My only remaining chance seems to be to secure a passenger immediately. My disappointment is most bitter because at last I seemed to have the chance sought last year."[41] Although one might be tempted to think it was personal, that perhaps sponsors were shying away from Jessie because of the trial, she wasn't by any means the only one in a similar position for this contest. Competition for funding was fierce and many other potential contestants were waging a similar battle. Consequently, a number of advertisements appeared in newspapers offering to carry paying passengers at fares ranging from £250 to £400. Some fliers even offered to name planes after cities that sponsored them or to have advertisements painted on the plane in order to raise money.[42] Jessie's contacts in the aviation industry even tried to persuade Edgar Percival to let her fly one of his Percival Gulls, the plane in which Jean Batten would make a record-breaking flight from England to Australia in 1936, but Jessie claimed he said he couldn't spare one due to economic hard times.[43] Cost was the major factor that in the end certainly deterred Australian women from participating in the race. "It is not because our girl aviators would not like to be competing, but the cost of such a world flight runs into a big sum," *The Australian Women's Weekly* lamented, quoting Lores Bonney, the first woman to fly in 1933 from Australia to England, as saying that "without finance for the specially-built machine it would be sheer madness to make the attempt."[44] Well, Jessie was eminently sensible and in the end, with no major financial support forthcoming, she pulled out of the race line-up. She wasn't the only one. Only twenty entrants finally took off for Australia on the morning of 20 October 1934, among whom were just two female pilots who were both eventually forced out, Johnson and Cochrane, along with a lone woman passenger. The winning entrant, the English Comet *Gloucester House*, flown by C.W.A. Scott and Tom Camp-

Eleven. England, My England

bell Black, touched down at Flemington racecourse in Melbourne 2 days 23 hours 18 seconds after leaving London.

Still, Jessie wasn't prepared to throw in the flying towel just yet, and so she set about obtaining her English private pilot's license at Hanworth Aerodrome, for which she had to do ten hours solo because she didn't have the money to pay for the extra flying time needed for a commercial license. In any case, she would have found it difficult to find work as an airline pilot because at that time they didn't usually employ women. She couldn't teach flying, either, because she'd never instructed before, so, she recalled later, "I thought the only thing was to plod on with the death or disaster flying, break records, go around the world backwards, or do something crazy."[45] Then a friend of hers, Colonel Jellico, offered her work in his office at Hanworth. He was currently planning a large aerial survey project in West Africa and he wanted Jessie to pilot the airplane from which the survey would be carried out. Before that could happen, though, there was a lot of paperwork to be done and maps to prepare and so Jessie would go into his office every day to help, and it was during the course of this preparation in late 1934 that she met the very remarkable and completely unstoppable the Honorable Mrs. Victor Bruce and her handsome, dashing business partner and pilot, Flight Lieutenant John Barnard Pugh, AFC.[46]

Although she'd only been flying four years, Mildred Mary Bruce had covered a lot of ground before that and most of it very quickly. The first woman to be convicted of a motoring offense in England, she'd married the winner of the 1926 Monte Carlo Rally, the Honorable Victor Austin Bruce, after which she'd placed sixth in the 1927 Rally herself. She held the joint record with her husband for having driven farther north than anyone else, some 270 miles beyond the Arctic Circle into Lapland, and in 1929 she captured the world twenty-four-hour solo driving endurance record. Just for good measure, she also held the cross–English Channel speed record and the twenty-four-hour distance record of 691 miles for powerboats in September 1929. In June 1930, she purchased a Blackburn Bluebird IV aircraft on a whim; she qualified for her license and between September 1930 and February 1931 became the first woman to fly around the world solo, although she didn't actually fly over any oceans but crossed those with her plane aboard a ship. Since then, she had attempted the in-flight refueled endurance record, qualified for her commercial "B" license and begun her own small air company, Luxury Air Tours. Johnnie Pugh had been her co-pilot in the endurance record attempt when on 11 August 1932, during the very time the Lancaster trial was taking place in America, they had set a flight endurance time of 54 hours 11 minutes on the third attempt in a three-engine Saro-Windover flying boat, *City of Portsmouth*, over Ipswich, refueling in flight from a tanker piloted by Mr. Bruce. Although they were unable to beat the world record of 23 days due to overheating engine oil, having planned to stay aloft for a month, they did break the British record of 50 hours 38 minutes.[47]

After a while, Jessie noticed that whenever Mrs. Bruce came by the Jellico office Johnnie Pugh was with her and he would stop to chat. As she was always on the lookout for some new innovation with which to be involved, Mrs. Bruce's latest means of transport was the autogyro, a predecessor to the helicopter. She wanted to be the first aviator to fly from England to Cape Town in one and so she'd purchased a Cierva C.30 in which to make the attempt, which would also break the record for the longest autogyro flight. As Jessie was engaged in making Jellico's Africa maps, Mrs. Bruce asked if she would make hers as well. The two women had a lot of experience in common and stories to share and they got to know each other quite well.[48] Mrs. Bruce's autogyro was a difficult aircraft to fly and she didn't feel confident flying it alone, so she was going to be escorted by another plane, a twin-engine, six-passenger De Havilland 84 Dragon, piloted by Johnnie Pugh. With him would be a wireless operator and mechanic, and Mrs. Bruce asked Jessie if she'd like to come along, seeing as how she'd drawn up all the maps and was familiar with the terrain.

They took off from Lympne Airport on 25 November 1934 and flew into France, where they were grounded in the Rhone Valley for days because of the weather. The dense fog was right down on the ground, and so the four of them passed the time playing endless games of bridge while they sat around talking, and Jessie and Johnnie got to know each other very well. Pugh was Irish, having been born in Dublin in 1908, and flying had been his life. He'd joined the Royal Air Force at twenty and was a Flying Officer in the Reserve by 1932, when he'd been awarded the Air Force Cross in the King's Birthday Honour's List, the same list in which Air Commodore Charles Kingsford Smith had received his Knights Bachelor for services to aviation in the Commonwealth of Australia.[49] The following year, Pugh became well-known for saving the lives of his two female passengers when the Fairey Fox plane he was piloting, owned by Mrs. Bruce, caught fire in mid-air while taking part in an aerial pageant at the Ford Aerodrome on 14 July. He'd just carried out the "Fox Dive," a high-speed dive at 250 miles per hour from which he would pull out at only 50 feet above the ground and climb back to 1,000 feet, when his fuel line fractured, spraying fuel onto the hot exhaust pipes. In an instant, the powerful 450hp Curtiss D-12 engine was on fire and Pugh was blinded as thick smoke as flames blew back into the cockpit. As he attempted to keep the plane in a controlled sideslip descent, Pugh and his passengers soon had to undo their safety belts and climb onto their seats, as the floor caught fire. Then, of course, Pugh could no longer control his speed and landed so hard the undercarriage collapsed on impact as the plane skidded thirty yards down the runway and flipped over. Pugh crawled out and then went back to the burning plane to pull out the two women. Miraculously, they all survived unscathed apart from a few bruises.[50]

Finally, the fog lifted sufficiently for Mrs. Bruce to attempt a takeoff, but after coming so far, she lost control and crash-landed at Nimes, and so the autogyro had to be sent back on the train while Jessie and Johnnie flew back to England in the Dragon. When they returned, the West African survey project was still having difficulty getting off the ground due to lack of money and so, knowing Jessie needed the work, Mrs. Bruce suggested she might think about managing her Commercial office at Heston Aerodrome instead.[51] Once the autogyro was repaired, Mrs. Bruce wanted to make another attempt to fly to the Cape, but by this time Johnnie was tied up with airline business, and so she asked Jessie if she could find another escort plane and fly it for her, as she didn't want to attempt the trip alone. Jessie agreed, deciding to buy a Robinson Redwing biplane because it could land at a very slow speed and because it had a large luggage compartment, a very useful addition to a small plane. Using what was left of her own money and borrowing some more from friends, Jessie bought a red-and-cream-painted Redwing Mark III (G-ABRL), had flight maps made, and then had some long-range tanks fitted because the standard one held only fifteen gallons.[52] Built by the Robinson Aircraft Company, the first Redwing model was test-flown in June 1930, before being released onto the market for £575. They were simple, but strong, wire-braced biplanes with a rigid wooden box frame, a wingspan of 30 feet 6 inches and a length of some 22 feet 3 inches. One of the main Redwing innovations was the single cockpit with two seats side-by-side instead of the more traditional in-line twin cockpits such as those in the Avro Avian. You could at least shout at the person alongside you in a Redwing rather than having to pass them a note or use a speaking tube, and this design also meant a saving on costs involved in instrument duplication and a big improvement in visibility for the pilot. The other innovation that appealed to Jessie was the large luggage compartment behind the cockpit accessed by doors in the fuselage. Like the Avian, the wings on this plane also folded back so it could be stored or transported easily. The widely spaced landing wheels were independently sprung with no axle, enabling the plane to land safely on rough ground with minimal risk of flipping over; even the 85hp Hornet engine sat on a shock-absorbing plate. With a top speed of 92 miles an hour and a cruising speed of 84, a Redwing could reach 1,000 feet in 90 seconds, giving it a nimble maneuverability.[53]

However, a few days before they were due to leave, Mrs. Bruce suddenly canceled the whole venture because of a speaking engagement, leaving Jessie with an expensive plane and no way to recoup the cost. She didn't know what to do. Although she could possibly pay back part of the finance money out of her own funds, she'd have to do something out of the ordinary to provide the rest. With an engine about the same size as the one in the Avro Avian in which she and Bill Lancaster had flown to Australia, it was already too slow compared to other current aircraft to consider attempting speed records. So, her only chance for a record would be to go where no one had gone, yet even then those places were becoming rarer by the month, so in late 1934 Jessie came up with a plan that in its own way would give her some closure for Bill. She developed an idea for a flight that would combine a record attempt with a business venture: she'd become the very first aerial commercial traveler to fly the route from England through northern and central Africa to Cape Town, stopping at trading outposts to show samples of goods and take orders. She would also be following the same route as Bill Lancaster had planned to fly, and so perhaps she might see or hear something along the way that would provide a clue to his disappearance.

She took off from Croydon in her Redwing with its load of goods samples on the morning of 4 January 1935, heading into blustery winter weather that deteriorated the farther she flew across France.[54] After three hours trying to wend her way through clouds that were right down on the hills, she had to admit she was lost and landed in a field to ask directions, whereupon she was promptly surrounded, dragged out of the cockpit and arrested as a spy! Without realizing it, she'd landed in the middle of a prohibited military area and, unable to speak the local dialect at all, Jessie realized the only chance she had was to make a run for it. Pushing the men aside, she shoveled handfuls of mud away from the wheels, swung the propeller, leaped into the cockpit as the motor caught with a roar and took off downwind and uphill to eventually land at Dijon in pouring rain as night fell. The next day, the bad weather followed her all the way down the Rhone Valley to Marseilles, where she found the mistral was not going to let her go anywhere. Jessie was forced to wait five days until she couldn't stand it any longer and decided to take off, regardless. She fought the wind all the way through the Pyrenees to Barcelona, from where she flew down the Spanish coast to Alicante, arriving there at seven in the morning to find the aerodrome completely deserted because of a near-fatal accident caused by treacherous downdrafts. Taking one horrified look at the bloody and broken pilot, Jessie decided to leave immediately before she lost her nerve. She made it, but the crosswind on the runway was so strong that a mechanic had to run alongside her Redwing plane as she took off, holding on to the wingtip. As she struggled over the Strait of Gibraltar, one wind gust was so violent it actually shattered the windshield.

Finally reaching Tangier, on the other side of the Strait, Jessie flew south to Fez and then turned northeast over the Atlas Mountains to Sidi Bel Abbes in northern Algeria before turning south again over the Saharan Atlas mountain range. After nearly five hours of wandering through valleys in search of passes through those rugged mountains, Jessie finally landed at Béchar on the northern edge of the Sahara, where the entire Foreign Legion outpost turned out to meet the plane only to be astonished when a woman climbed out of the cockpit. Although Jessie couldn't speak French and they couldn't speak English, there was a lot of good-natured consultation with language dictionaries as they worked out how to communicate. The Legionnaires set her up in a comfortable room in their guesthouse, after which she dined with the Commandant. "Before long," she wrote later, "I began to wonder if the French Air Force was stationed there specially to look after British airwomen."[55] Before she left that remote outpost on the northern edge of the Sahara, the men made sure she had adequate desert rations, which included, as well as a three-gallon container of water, some pâté de foie gras. Well, it was a French outpost. They also gave her strict instructions not to fly at night and that if she lost sight of the single track across the desert she'd be following during the day, the same one that Lancaster was supposed to have been following, she was to land immediately, get comfortable in the cockpit and go to sleep. If she failed to come out the other side, the Commandant assured her, he would fly out himself to look for her and he would find her.[56]

Jessie hadn't been in the air long before she realized the wisdom of his words. There was absolutely nothing beneath her but sand. For hundreds and hundreds of square miles in all directions, there were no landmarks at all, no people, no animals, not even a wisp of smoke, nothing except desert and clear blue sky. On the first night of her crossing, she landed at a well right in the heart of the desert known simply as Bidon 5. The next morning, she could only persuade the engine to start on one magneto. For five hours she flew with a stuttering engine, talking to it, cajoling it constantly to keep her in the air. As if that weren't enough, the windblown sand frequently obliterated the track below her, leaving her to fly on relying solely on her compass as if she were crossing an ocean, slipping from side to side of the track's estimated position, hoping she'd pick it up again before it had curved away somewhere. One can only imagine what must have gone through her mind as the hours went by under that relentless sun. Perhaps she was flying over the very place where Lancaster had gone down, over the remains of him and his plane, and she may well have kept watching for any small sign of her friend that searchers may have missed. Finally, on 24 January she reached Gao, the destination that Bill had never reached, situated on the upper reaches of the Niger River in what was then French Sudan, about 400 miles downriver from Timbuktu. When she finally sighted the Niger River after so much desert, Jessie was so relieved, "I simply yelled for joy." Gao's population at that time was probably about 5,000. It had always been a terminus for river traffic, for from here southwest the Niger is navigable for over a thousand miles until it empties into the Gulf of Guinea, but in the mid-thirties Gao was rapidly growing in importance as the terminus hub for trans-Saharan road and air travel. Only eighteen months later, Amelia Earhart would use Gao as a major refueling point on her attempt to fly around the world.

Most of the desert was now behind Jessie, and as she followed the Niger southwest the landscape beneath her changed noticeably as scrub replaced sand. Now she regularly flew over native villages that from the air looked to her like "enormous clumps of mud-covered mushrooms." She flew over the military post at Niamey and then turned farther west into Benin to land at Parakou, from where she would fly on to Cotonou in Benin. As she passed over Gaya, she noticed the topography becoming quite hilly and the foliage denser. Even in the air, the cockpit became like a sauna in the intense heat, and her visibility deteriorated as she groped through dense clouds of smoke from incessant bushfires. She had been warned to be careful when landing at Parakou because of treacherous crosswinds and a hill in the middle of the airfield, and as Jessie circled the field for the third time she could see the wind sock changing direction constantly, but she still made one of her best landings and taxied in to refuel.

By now it was around noon, and even in the short time she was on the ground the propeller blade had become so hot Jessie burned her hand trying to spin it when she went to restart the plane. "The best slimming exercise I know," she later reflected, "is to swing one's own propeller in an African sun." With little "lift" in the air because of the heat and a short runway, it was a much more hazardous takeoff, and it stayed hot as she flew towards the African coast. From that altitude, the whole of Africa seemed to be on fire, she thought, as she fought constantly with thermal updrafts and pockets that would then drop her a thousand feet without warning. She'd regularly lose sight of the ground in the dense smoke and consequently nearly flew into hills at least twice. After three hours of this, her arms and legs ached terribly, but as she neared Cotonou she started to sing loudly at the welcome prospect of rest.

But her relief was short-lived. Suddenly she was battling a stiff headwind, anxiously watching a petrol gauge veering towards "Empty" as she revved her engine hard to keep her nose up. Then, twenty miles out, she was enveloped in a dense fog bank that blotted everything out. Without the fuel to go up, she had to let the plane down while flying totally blind, hoping desperately that the fog wasn't sitting completely on the ground and there would be clear air between the fog and the trees. With only 150 feet left, she found the clear space and flew on, skimming the treetops, seeing only swamp and no landing field. She was tempted to land on the grass until she took a closer look and realized it was eight feet high and growing in water. The petrol gauge showed "Empty." Jessie was rarely scared when flying, but she was now horribly frightened. Looking behind her through the trees, she realized she'd flown over a village. Turning back, dropping lower, she could see there was a main street, but the villagers who rushed out to see her just looked on in amazement instead of pointing to any likely landing area. Her gauge had been registering empty for the last ten minutes, and Jessie was expecting the engine to splutter to a stop any second. She'd run out of options; her only choice left was to land on the road beneath her. Picking out a likely deserted section of it, she slipped between the trees and touched down on the track safely, but while still rolling down it at some 40 miles an hour Jessie saw to her horror some people run out in front of her. With the nearest only a few meters away and making no effort to avoid her, Jessie had to swerve hard left, and the wingtip clipped a tree. With a loud crash, her plane sideslipped into a mud bank on the side of the road.

As the sound died away and the dust settled, there was a terrible silence. Jessie crawled out to see the individual she'd wrecked her aircraft to save still standing in the road, petrified with fear. When Jessie called for help, he ran away. She swore a lot and then lit a cigarette as she surveyed the damage. She quickly concluded her situation was hopeless. Nine hours of flying that day had come to this. The Redwing had hit the bank so hard it had broken in half; the cockpit was entirely destroyed. She had no idea how she was still alive. By now, every African for some distance around had gathered to see the bird fallen from the sky. Most of them had never seen an airplane until now. Using a lot of sign language, Jessie discovered there was an Englishman living in the village but that he wasn't due back there until evening. Just then, three English people appeared. They'd been waiting at the Cotonou airfield for her, and after hearing that she'd come down here only a few miles east at Abomey-Calavi instead they'd rushed back to find her. Jessie was driven back to Cotonou, where she stayed for the next three weeks until the next boat called in, for in that part of the world the damage to the Redwing was irreparable. While she was waiting there, one of the Englishmen invited her to go with him on a business trip farther inland to Bohicon and then to Abomey to see the legendary palace of the last native king of Dahomey, now the Republic of Benin.[57] Abomey was the capital of the old Kingdom, which had been a major center of slave trading. The thrones of the palace were supported by the skulls of defeated chiefs and the mud of King Glele's tomb mound was mixed with the blood of human sacrifices. "The whole place smelled of death," Jessie wrote, "and I was glad to get back into the sunshine." The night Jessie was taken to see a fetish dance held where "great trees interlaced overhead and the moon shone through the foliage.... Native torches were burning and in the reddish glare I saw hundreds of natives dressed in barbaric costumes wildly dancing to the ceaseless beat of tom-toms."[58] After a few hours of this, a chief called for silence and spoke at length to the crowd, recounting the story of Jessie's arrival in a big silver bird from the sky. By then it was nearly dawn, so as the dancing resumed Jessie unobtrusively slipped away, and they left for the return trip to Cotonou. A few days later Jessie boarded her ship for England, where she arrived with vivid memories of West Africa but once more in debt and bitterly disappointed. Unknown to her, however, she had achieved one particular record: her plane had traveled farther than any other Robin Redwing would ever fly.

Having no money left to fly and with debts to pay off now, Jessie took up Mrs. Bruce's offer to be the manager at Heston Aerodrome for her Commercial Air Hire Company. Bruce had established the company in August 1934, to make newspaper, freight and passenger flights between London and Paris, and in April 1935 they had begun passenger air shuttle services between the Croydon and Heston airports. Heston was to the west of London in Hounslow, Middlesex, and a few years later it would become part of history as the airport to which Prime Minister Neville Chamberlain returned in 1938 from his "peace in our time" meeting with Hitler. So, this was a very responsible position in a rapidly expanding company, rather more than the typist's job that newspaper reporters credited her with when they tracked Jessie down there in December. Still, grounded behind a desk of any name, she was like a racing pigeon with its wings clipped. "I'd give anything to get into the air again," she lamented, "but what can you do when there's no money. I'm broke and still trying to pay off my debts.... I used to make £2000 (around $US10,000) a year on average out of flying then. Now—well, there isn't much left. You see, I never saved any money. You don't, somehow, when everything is rosy. It doesn't occur to you that it will ever end ..." She talked about her disastrous trip to Africa, which had left her with a lot of debts to pay off on only a small salary, but then declared, "Somehow I'll win through and pay them, and then I'll try and fly again. Ordinary record flying is finished. There is work and money, though, for anybody with ideas. I'd like to do some aerial commercial travelling, flying to out-of-the-way places with goods the people there never see. Or I could demonstrate new types of airplanes as I did in America. And I believe I'm the only woman who knows how to handle a flying boat." She'd returned from America with a terrible inferiority complex, she explained, and had felt useless and unwanted. But she'd made some real friends where she was working now and was feeling more confident that "when I'm free of debt, I'll fly again, although there isn't a great deal of a chance for a woman in this country—or there doesn't appear to be much, but there must be something and I'll find it. That's all I'm waiting and longing for, to get back to the air ... the air." She spoke of the air, the journalist observed, as another woman, a lesser woman, might speak of the man she loved.[59]

But sometimes life opens up another runway. Within a few months, Jessie could speak of the man she loved, none other than Johnnie Pugh, still a flight lieutenant in the RAF Reserve, now a co-director with Mrs. Bruce of the Commercial Air Hire Company and a very busy pilot. Among his other work, he was in command of two daily flights from the Duxford airport in Cambridge that collected forecasting weather data for the British Meteorological Office in London. For these flights, commonly known as the "Met. Flight," Pugh had to fly at an altitude of some 30,000 feet, which in the mid–1930s was like going to the moon. Dressed in an electrically heated high-altitude suit while seated in an open cockpit, he would take and record readings of the air temperature, humidity and air pressure every 1,000 feet. At that altitude, ice would frequently form on his goggles and oxygen mask and he'd often feel sleepy and find it very difficult to concentrate because of the altitude. After taking the last observation, he'd then descend in a spiral to avoid frostbite or the engine freezing up if he lost height too quickly. By then, too, in a plane not even equipped with a radio flying thousands of feet above the cloud layer, Pugh really wouldn't know accurately where he would come out below the layer. In effect, he'd been drifting up there on air currents for ninety minutes, so when he came out of the spiral at 9,000 feet it was always something of an adventure to discover his location. If there was too much cloud, though, he'd sometimes be down to 1,000 feet or less before knowing if he was over land or sea! By August 1935, he'd made some 8,000 of these high-altitude flights, putting him in a pilot category of his very own.[60] Then later that year he and his wireless operator, R.F. Burgess, had been lucky to be rescued by a French fishing boat from the middle of the English Channel when Pugh's Commercial Air Hire Company mail plane went into the water as a result of engine trouble as they were returning from Paris. After hearing the radioed "Mayday" distress call, a large group of mail and passenger planes, as well as a lifeboat, had carried out a fruitless extensive search in rough seas and had almost given up when the fishing boat found them, suffering badly from hypothermia.[61]

Three months after despairing that all she wanted to do was get back in the air, Jessie was declaring that there would be "no more flying stunts for me" a few weeks after she and Pugh announced their engagement on 9 March. "My wild days are over," she announced to reporters. "I've had my fun. I'm just going to sit back and let John do the piloting for both of us now." However, she did admit that home life would never agree with and her and that she'd always have to be near airplanes so, with John's wise approval, she was going to keep on working with them. "It's taken me a long time to learn sense," she admitted, "but I think I can see things in their proper perspective now."[62] At last, Jessie had fallen in love with someone who was her equal, with whom she had much in common, and who was willing and able to settle into a long-term married relationship with her, and she wasted no time in saying, "Yes." She and John, who was now a pilot with British Airways, were married on 16 May at the Registry Office in Epsom.

Marriage didn't change Jessie's opinions about women and flying, though. With rumors of war on the horizon, she gathered the support of a number of leading British airwomen to form a corps of pilots if conflict broke out who would undertake active flying duties, other than combat flying, such as piloting aerial ambulances, carrying dispatches and training pilots. She clearly saw that "if another war breaks out we shall need all available pilots, both men and women." They could also take over commercial airline piloting to free up the male pilots for active frontline duty. She proposed that old military aircraft should be reserved for training purposes instead of being scrapped, so that women could be trained under air force conditions at air force bases. "Aviation experts are agreed," she pointed out emphatically, "that in the piloting of aircraft women are little inferior to men."[63] As she must have known it would, her proposals revealed not a little misogyny among the male pilot ranks, of which a certain Captain H. Duncan Davis was a prime example. "Nothing would ever persuade me to take part in war-time aviation with women pilots," he harrumphed. "Generally speaking, women are just as unreliable in the air as they are on the roads."[64] A reply by well-known New Zealand airman Francis Chichester summed up the argument in spirit, really, even if not entirely accurately. "Women make better air pilots than men," he declared. "The proof of it is plain. Women pilots are all alive."[65] Jessie must have felt a glow of justification when Commander Pauline Gower MBE was given the job of forming the women's section of the Air Transport Auxiliary in 1939. Eventually, 166 women from all over the world volunteered to join during World War II, of whom fifteen lost their lives, including Jessie's friend, Amy Johnson. From 1943 they earned the same pay as men of equal rank in the Air Transport Auxiliary, the first time the British government agreed to equal pay for equal work within an organization under its jurisdiction, and they flew virtually every aircraft flown by the RAF, including four-engine bombers.

Probably about the time John joined British Airways, he and Jessie moved to Singapore, where the Australian aviatrix Nancy Bird Walton met them in July 1938, while visiting various countries to investigate aviation development there. Walton felt a close affinity with Jessie, not only because they had both been pioneering airwomen in their own way but also because Walton was only the third Australian after Jessie to be a member of the Ninety-Nines and because she may well have been the only other woman to have flown the *Red Rose*. In 1934, she hired the *Red Rose* for a training flight from Sydney to King Island in Bass Strait. With Cliff Carpenter as her instructor, she was flying in formation with pilot Charles Gatenby when they became lost in low clouds over the Strait. Unable to locate King Island, they finally landed safely on the beach at Apollo Bay in Victoria. Following in Jessie's footsteps, Walton took part in three of the modern Powder Puff Derby race series that took place between 1947 and 1977. In 1958, she was the first overseas entrant in that series and with her American co-pilot Iris Critchell placed fifth.[66]

Jessie and John returned to England at the outbreak of World War II. In July 1940, John relinquished his RAF commission for one in the Royal Naval Volunteer Reserve, becoming Flight Lieutenant Pugh 27014. During the war, he commanded a series a series of Fleet Air Arm bases, including Royal Naval Air Station Stretton (HMS *Blackcap*) in Cheshire, the Naval Instrument Flying School at RNAS Donibristle (HMS *Merlin*), and the Naval Advanced Instrument Flying School at RNAS Hinstock (HMS *Godwit*) in Shropshire, which was the main training facility for "blind" flying using a transmitted radio beam as a homing signal. Using what was known as the Lorenz Beam Approach, a pilot was guided back to base by flying along the path of a continuously transmitted single note signal. If the pilot deviated to one side, the signal broke up into short beeps or "dots"; if they deviated to the other side, it broke up into longer short notes or "dashes." This guidance system was exceptionally valuable in situations where bad weather, crew fatalities or aircraft damage rendered locating a base problematic and so it saved many lives, and establishing this training base against substantial opposition was Pugh's own successful mission. In recognition of his services, he was appointed as an Officer to the Military Division of the Order of the British Empire in 1944.[67]

After the war was over, Pugh flew the first British Empire Airways service into Southampton and then in 1947 he became Chief Pilot for Cyprus Airways, an associate of BEA. In 1949, he was promoted to Senior Captain, First Class, and in June 1955, still with Cyprus Airways, he was awarded the Queen's Commendation for Valuable Service in the Air (Silver Oak Leaves) as part of the Queen's Birthday Honours.[68] Jessie must have been very proud of him, but in her case she had been right: for her, there were no more stunts. They bought another house in Spain and down-sized to an apartment in London, but it meant Jessie had to sell her beloved baby grand piano she'd had since she lived in New York. She'd always been passionately fond of music, hinting that if she hadn't taken up flying she would have been a concert pianist. She must have thought that at last she had faded from sight, that as she got older life had finally settled into peace and quiet. But life had one more surprise.

Epilogue: Full Circle

One morning in late February 1962, Jessie had gone down to the kitchen to make breakfast when the telephone rang. She was busy heating milk and so John answered it. She heard him say, "Well, that's very strange."

"So, what have you been doing?" he asked, coming back into the kitchen.

"Nothing," Jessie protested. "What do you mean?"

John replied that a friend of theirs had just called to say Jessie was on the front page of the *Daily Mail*. John and Jesse only had the *Daily Express*, and they hadn't looked at it yet.

Neither of them could think of anything Jessie had done lately that would get her into a newspaper.

"Have you been pinched by the police for speeding or anything?" asked John.

"No," Jessie replied. "I've been perfectly good! I haven't been up a one-way street or had any parking complaints. I can't make out what it could possibly be. I wonder what she's talking about?"

It was only then that her husband thought to look at the front page of their *Daily Express*.

"My God!" he exclaimed from the dining room. "Here it is!"

Jessie brought their breakfast in, and as she sat down he laid the newspaper out in front of them on the table. "Well," he said slowly, "there you are."

With mounting shock, they began to read what had happened in a far corner of the Sahara desert a few days ago.[1]

On the morning of 12 February, three heavy trucks of a motorized platoon of the Group Saharien Mixte du Touat were carrying out a reconnaissance in the area known as the Tanezrouft in the very heart of the Sahara, about 100 miles west of Bidon Cinq on the trans-Saharan motor track, the very place where Jessie had landed on her flight across the Sahara. A desert within a desert, the plain is so arid, flat, featureless, trackless and bereft of any life that it's known as the "Land of Thirst" by the nomadic tribes who shun it. It is so hot and dry there that a person requires a gallon of water a day to survive. Attempting to negotiate around an impassable area of soft, dry sand between them and the track, the patrol had turned east onto hard sand for about a hundred kilometers when they saw a tiny object in the distance that was hard to define. It wasn't until the convoy had almost reached it that they realized it was a crashed aircraft!

As this had been an area near where French atomic bomb testing had taken place, the patrol adjutant cautiously halted the convoy a short distance off and walked to the wreck alone. Standing there on that vast, stony plain, he realized he was looking at the skeletal remains of a small, single-engine biplane lying on its back. The fabric covering had been stripped from the frame. The nose of the plane and its broken propeller blades pointed east, towards the distant track. Then the adjutant noticed what he had been dreading: under what had been the starboard wing, he could see a human skull. Stepping closer, he could see that in fact there was a mummified body lying there on its left side, partly covered by a shallow drift of sand, and as he bent down and gently blew away the sand a scar became clearly visible above the right eyebrow. Standing up again and looking around, he saw with amazement that the pilot had tied his belongings to the wing of the plane above him and they were still there—completely intact! A thin metal case contained his passport and other documents and a large waterproof envelope inside which was his logbook. The last entry had been scrawled on 12 April 1933, and inscribed on the cover was the name Captain William Lancaster.[2]

Jessie and John sat back and just looked at each other in silence. "It was a shock, as you can imagine," Jessie recalled later. "A most appalling shock … for both of us. I didn't know what to think. What can one think? It was a most colossal shock that anyone can imagine."[3] It was as if Bill had suddenly risen from the grave in front of their eyes. They had been married for twenty-six years now, and Jessie's early life in the clouds was now memories. But more was to come. Within the pages of Lancaster's logbook was his diary written during the eight days that he waited to perish of thirst in that sandy hell. Having had some experience of Bill as a diarist, Jessie could imagine what might be within its pages and she was immediately anxious. She had another life now. The last thing she needed was her previous one being dredged up again, but she could already see it coming. According to the terms of Lancaster's will, she was the sole beneficiary of his effects and so these last possessions of his were now hers. The past had just reached out and taken hold of the present, and after all this time she would know Bill Lancaster's thoughts at the end of his life.

Needless to say, newspaper reporters were immediately on the telephone, but Jessie refused to be interviewed, referring all inquiries to her solicitor, including lucrative financial offers for publication of her story. It wasn't just the romance angle, of course; here was the last chance Lancaster might have answered the one question that had hung over Jessie's head so many years like an invisible sword of Damocles: Did he shoot Haden Clarke? The sensationalism value was enormous and, Kiki Lancaster having died in 1953, Lancaster's two daughters promptly flew to Algiers in an attempt to gain possession of the documents. However, the British Consul-General stood by the wording of the will and, under instructions from the Foreign Office he refused to release Lancaster's possessions to anyone but Jessie. So, after some delay because of the Algerian civil war, Jessie accepted Bill's belongings and in this way his final words came to rest in the hands of the woman to whom they had been written.[4] Jessie and John read them together, and when they finished John, deeply moved, had tears in his eyes. Here was a fellow pilot who had courageously faced death utterly alone with not a whimper or grumble, knowing there was no hope of rescue, knowing virtually the day of his death. "He was a very brave man and one I'd have been proud to call my friend," he whispered. "This is such a brave bloke I really feel that it is only fair to him his story should be told. That is courage personified."[5]

And so, in October, they published Bill Lancaster's last words in full in the *Daily Express*, a newspaper with whom Jessie had a long, collaborative and trusted relationship for many years, accompanied by a foreword she wrote that began:

> I had never forgotten Bill Lancaster. The world we had known together, the roaring twenties, the death-or-glory record flights in tiny biplanes, the Depression when there wasn't much in the way of pickings for pilots like us, then drama, headlines, and Bill's tragic exit from it all; it was half a lifetime away. The passing years had taken the sharp edges off the memories. Sometimes it seemed like a different world. But it hadn't been another world. The headlines that said his body had been found told me that.

It had been a shock, she went on, to realize that Bill hadn't died suddenly in some flaming plane crash but had perished slowly over a week of thirst, stranded, immobilized, lost in the desert, ultimately realizing there would be no rescue. Even so, he had met death courageously, with "no whimpering, no recriminations for him." She and Johnnie wanted this diary "to be a memorial to a man who died as a hero," she declared. "To all who read them, these last words will be an inspiration."[6]

His logbook entries revealed that Bill had been forced down during the night of 12 April on his way to Gao when the engine spluttered and cut out without warning. There was no moonlight and so he'd not been able to see the ground to judge his landing speed. When his wheels had broken through the surface crust and dug in, the plane had flipped over. Bill had received a nasty head wound, but he survived, although when he reviewed his situation in the morning he might have wished he hadn't. He was carrying no food, only about two gallons of water that he calculated might last a week with careful rationing. "It will give me time to reflect and write a few notes in this log book," he began. He and Jessie had discussed emergency tactics before he'd left and they had decided if he came down he should stay with the plane and so, knowing that search patrols would be sent out and that Jessie would come looking for him if she could find a plane, the wounded Bill methodically prepared material for lighting fires and flares at night so he would be seen. However, in the darkness of the night he had drifted away from the trans-Saharan motor track that he was supposed to be following as a guide, and now no one knew his actual location and so, aware the probabilities were not in his favor, Bill also made plans for securing his possessions and documents in case he wasn't found. Once his water was gone, he knew he would perish within hours in that heat.

So, that first day as the French began searching along the track and north of Gao, Bill calculated that in order to hold out as long as possible he'd have to ration water to one pint a day, which would not allow him to clean his wounds from the crash, blood loss from which was already weakening him. "Truly," he wrote, "I am atoning for any wrong done on this earth."[7] He lit flares at night, not realizing he was much farther away from the track than he thought, and ended that day in hope: "What a strange and powerful thing is faith. I have faith that someone will come and find me." But no one came and on the next day, Good Friday, he began to focus on conserving water, flares and matches and sheltering from the heat. If it came down to the end, he inscribed with stiff upper lip, "Please God I pass out like a gentleman."[8] Unless a miracle happened, Bill Lancaster now knew the day of his death.

Bill declared his love equally for his mother and Jessie every day, referring to them both as "my darling," and placed a lot of hope in Jessie organizing a search. All these years later, it must have caused Jessie pain all over again to remember her desperate attempts to do just that in the face of refusals to help. By Sunday, Bill was beginning to suffer badly from hunger and thirst. His only remaining food was one chocolate bar and he was constantly battling his desire to drink water: calculating that he had only three days' supply left at the most, he dreamed of running into his mother's kitchen and turning on the tap into his mouth. He comforted Jessie and his mother not to grieve for him, because "I have only myself to blame for everything. That foolish headstrong self of me." By Monday, he had used the last flare to no avail and, although still professing hope, knew his water supply would last only one more day.[9] By that evening, he was becoming resigned to his fate and reminded Jessie he'd stuck to their agreement and stayed with the plane. He asked his mother to kiss his two daughters and to make sure Jessie received the logbook, and he asked her to tell his wife, "She can now really forget." He was growing weaker and could only write a few words, but by Wednesday, his seventh day, he knew he was down to the end of his water supply, exactly when he'd originally calculated, and so he made himself comfortable and settled down to write his last message to Jessie. "Give up flying," he advised. "You won't make any money out of it now. Collect what money there is available. Pay back my mother and my father what is just and then book passage for Australia.... You will always think of 'Old Bill' as a good scout. Too bad I had to go like this—think of me occasionally and write your book. I'd like to think it would be dedicated to me." He said good-bye to his family and to his mother and father, making it clear that he wished for his mother and Jessie to reconcile. "Forgive your enemies should be included in your password," he lectured his mother. Then he secured everything in the waterproof bag and hung it in clear view where it could be found. The next morning, he wrote a final few words on a fuel card: "So the beginning of the eight day has dawned.... I have no water. No wind. I am waiting patiently. Come soon please."[10] Then he lay down under the wing to die. Three days later, the search was called off.

Jessie could rest easy: Bill didn't write anything about that tragic night in 1932. Those hoping for a last word on that matter would be forever disappointed. Bill Lancaster, who died next to an airplane originally named the *Southern Cross Minor* while declaring his love for a woman originally from the town of Southern Cross, was buried in the civilian cemetery at Reggane where he still lies today, he and his secret secure in the desert sand. "He was always getting into some kind of trouble," Jessie remembered.

> He used to do mad things without thinking. That is why we always used to say our combination was good because I had a Scottish mother and I always used to say my Scots caution stopped his Irish madness. We used to have frightful rows over flying. In fact, I once hit him over the head with a spanner saying, "Will you stop doing those damn fool crazy tricks with this airplane!" He used to adore to aerobat. Everything would fall out and I'd nearly fall out, but I curbed him because I had horse sense. I knew all the best and worst about him. We were such good friends. We had great fun together.[11]

Nine years after Lancaster's body was found, Jessie Miller, the first woman to travel by airplane from England to Australia, the first woman to fly over the equator, the first woman to fly to so many places, made her final flight into the blue.

Jessie never did publish an account of her life and so, finally, she has been able to tell *her* story here.

Jessie Keith Miller was a pilot once...

Chapter Notes

Prologue

1. See Eric Tagliacozzo, *The Longest Journey: Southeast Asians and the Pilgrimage to Mecca* (New York: Oxford University Press, 2013), 115–116. Also Cedric Watts, *A Preface to Conrad*, 2nd ed. (New York: Routledge, 2014), 22–24. In a few years' time, however, Bankga Island would gain a reputation of an entirely different kind as the site of the notorious massacre by soldiers of the Imperial Japanese Army of twenty-one uniformed Australian nurses and a number of wounded. See Betty Jeffrey, *White Coolies* (Sydney: Eden Paperbacks, 1954).

2. For their arrival in Singapore, see "The Red Rose Arrives," *Straits Times*, 7 January 1928, 9.

3. This prologue is based on Jessie's own account: "The Red Rose Crashes," *The Register* [Adelaide], 28 April 1928, 11.

Chapter One

1. Established as the Swan River Colony in 1829, Western Australia was renamed in 1832 and became a state of Australia in 1901.

2. Crux, or the Southern Cross as it is more usually known, was first cataloged as a separate constellation in 1603. As it orients to the South Celestial Pole, it is as significant for celestial navigation in the Southern Hemisphere as Polaris is in the Northern Hemisphere. Because of its prominent visibility in the Southern Hemisphere's night sky, it appears on the flag of Australia as well as those of a number of Pacific nations. It also features in the words of the Australian national anthem, "Advance Australia Fair," written by Peter Dodds McCormick, in which Australia is situated "beneath our Radiant Southern Cross."

3. In 1895, Jessie's father was an accountant at the Commercial Bank in Sale, Victoria, when the manager, Arthur Short, became very depressed and neglected the balancing of the Bank's daily figures. When Short heard in November that he was about to be audited, he murdered his family and shot himself. (See "The Inquest," *Gippsland Times*, 21 November 1895, 3.) The associated trauma of an event like that could explain why Charles moved about as far from that place as he possibly could while remaining in the same country.

4. On Lot 36, currently occupied by the ANZ bank building. This history of Southern Cross is drawn from information provided by the Southern Cross Museum and Historical Society, including the history of banks in the town compiled by Lance Stevens at the Society.

5. The British India Mail steamship *Cashmere* was an iron-hulled steamship of 1,015 tons launched in 1864. On Thursday evening, 5 July 1877, she left Zanzibar bound for Aden but ran onto rocks near Cape Guardafui. The crew were attempting to launch the boats during the night when the heavy seas destroyed one that contained mostly women and children, drowning seven people, including the Beveridge family and their servant. The survivors were stranded on the bleak shore for five days, The eighty-nine survivors eventually walked and then sailed in small boats to Abola, from where they were rescued by the *Queen Margaret* on 10 July, finally reaching Aden two days later.

6. Also a legendary genealogist, a Fellow

237

of the Royal Genealogical Society, Sidney A. Beveridge was the author of the impressive *Story of the Beveridge Families of England and Scotland* (Melbourne: McLaren, 1923), from which much of this family background is drawn.

7. For more of Tom Oates' story, see Trish Burgess' *Bearing the Heart of a Sailor: Letters from Antarctica and Other Faraway Places from Tom Oates to Elizabeth Eadie—1936–1938* (Fisher, ACT: 2004).

8. The Rev. John Kay Hall, vicar of Holy Trinity Church, Wangaratta, Victoria, had six sons and three daughters. His wife, Jessie, who died in 1910, had arrived in Victoria in 1838. In 1888 he was minister of the Church of England in Gisborne.

9. The largest subdivision is actually the Sakha Republic, the capital of which is Yakutsk, within Russia. At 1,198,152 square miles, it's approximately the size of India.

10. "Trinity College Examinations," *Barrier Miner* [Broken Hill], 26 August 1913, 2. Also, "Convent of Mercy, Broken Hill," *The Catholic Press*, 8 January 1914, 36.

11. "Former Resident of Timaru," *The Press* [Canterbury, NZ], 4 December 1930, 11.

12. Comments on marriage from Ralph Barker, "Mrs. Pugh," transcripts of interviews conducted c. 1968, Tape 3: 8. Also "Had Other Ideas: No Home and Fireside," *Cairns Post*, 27 June 1931, 5.

13. Barker, "Mrs. Pugh," Tape 3: 8, 17.

14. Mike Walker, *Powder Puff Derby: Petticoat Pilots and Flying Flappers* (Chichester: Wiley & Sons, 2003), 163.

15. Barker, "Mrs. Pugh," Tape 3: 18.

16. Donald Mackenzie, "Chubby Miller: Her Rise to Fame," *Maitland Daily Mercury*, 29 January 1931, 10. Originally from the Australian state of Victoria, Mackenzie wrote for newspapers all over the world, including as a war correspondent for the New York *Daily News* until his death in 1949.

17. Inspired by her friend, Margaret Starr also decided to break records by returning to Australia adventurously accompanied by an ex–RAF pilot—by car. In February 1928, she announced that she and Captain F. Robinson would depart England in March in a two-cylinder Jowett. They would drive across Europe into Italy, board a ship for Port Said, drive through Africa to Cape Town, take another ship to Perth and then resume driving to Sydney. They were reported to have left in July but may not have finally reached Australia. (See "Out for Adventure," *The News* [Adelaide], 10 February 1928, 7.)

18. Jessie's version of their meeting in the Barker, "Chubby Pugh," Tape 1: 1.

19. Ross and Keith Macpherson Smith, later both knighted, were brother aviators who were the first pilots to fly from England to Australia, in 1919. Smith and Bennet were killed in 1922 testing a Vickers Viking IV amphibian in England.

20. Parer would later become an aviation pioneer in New Guinea, reputedly being the first pilot to fly over the Owen Stanley Range.

21. A former test pilot for de Havilland, Cobham had made a 5,000-mile air tour of Europe in 1921. In 1928 he flew a Short Singapore biplane flying boat around the continent of Africa. One of the founding directors of Airspeed Limited, he carried out some of the early experiments with in-flight refueling.

22. Information on Bill Lancaster's background from Ralph Barker, *Verdict on a Lost Flyer: The Story of Bill Lancaster and Chubbie Miller* (Sydney: Fontana/Collins, 1986).

23. Barker, "Mrs. Pugh," Tape 3: 8.

24. Barker, *Verdict*, 24.

25. *Ibid.*

26. Barker, "Mrs. Pugh," Tape 3: 7.

27. *Ibid.*, 18.

28. *Ibid.*, 12.

29. *Ibid.*, 27.

30. Barker, *Verdict*, 26–27, and Barker, "Mrs. Pugh," Tape 3: 13.

31. "'Broncho' Lancaster to Fly to Aussie," *Northern Standard* [Darwin], 11 October 1927, 1.

32. Mrs. Keith Miller, "Only One Frock for Half a World's Journey," *The Mercury* [Hobart], 18 April 1928, 7.

33. He would die only three years later, along with the Air Minister, Lord Thomason, in the crash of the British airship R101 in France.

34. Barker, "Chubbie Pugh," Tape 1: 2.

35. Barker, *Verdict*, 25.

36. Barker, "Mrs. Pugh," Tape 3: 9.

37. Barker, "Chubbie Pugh," Tape 1, 1.

38. Bill would later register *Red Rose* in Australia in 1928 as G-AUTU.

39. These details are from the A V Roe & Co. Ltd. production records, which also reveal that this *was* the first Avro 594 Avian

III. Its production code R3/AV/125 is prior to R3/AV/412, which was Lady Heath's Avro 594 Avian III, registered G-EBUG [C of R 1485], which has also been claimed to be the first. That plane's Certificate of Airworthiness 1245 was not issued until 27 November 1927, and it was not registered [1510] to Lady Heath until 29 October, before being registered to Amelia Earhart as NC7083 in September 1928.

40. Sir Charles Cheers Wakefield, 1st Viscount Wakefield, had founded the Wakefield Oil Company, later renamed Castrol, a name derived from the castor oil used as an additive in the company product. As well as supporting Jessie and Bill, Wakefield also later sponsored New Zealand aviatrix Jean Batten to purchase the Gypsy Moth in which she surpassed Amy Johnson's solo record from England to Australia and then the Percival Gull Six in which she set a solo world record from England to Brazil in 1935 and the first solo flight from England to New Zealand in 1936.

41. Barker, "Mrs. Pugh," Tape 3: 26.

42. Charles P. Johnson, "Woman 'Stole' Plane for Solo," *Pittsburgh Press*, 18 November 1930, 34.

43. "To Australia by Air," *The News* [Adelaide], 12 October 1927, 9.

44. Barker, *Verdict*, 28.

45. Prudence Black, "Fashion Takes Flight: Amy Johnson, Schiaparelli and Australian Modernism," *Hecate* 35 (2009): 68.

Chapter Two

1. Jessie would actually meet Elder two years later while competing in the 1929 National Women's Air Derby. They would both become charter members of the Ninety-Nines. Though forced to ditch her plane due to mechanical problems only 360 miles from land, Elder established a new over-water endurance flight record of 2,623 miles. At that time the longest flight made by a woman, it was almost immediately eclipsed by Jessie's record.

2. Frances Grayson had made her first transatlantic attempt from Maine with pilot Wilmer Stultz on 17 October in their Sikorsky S-36 amphibian, the *Dawn*, only to return because of bad weather. They made a second attempt on 23 October but returned again after only 500 miles. Stultz wisely took the hint and quit, but Grayson hired another crew and took off 23 December from New York for England via Newfoundland. They were never seen again. Ironically, Stultz lived on to pilot the Fokker trimotor *Friendship* that on 19 June 1928 finally did carry the first woman across the Atlantic in the air: Amelia Earhart.

3. Sir Charles Kingsford Smith is probably the best-known of Australia's pioneer aviators. A former World War I pilot, he gained his commercial license in 1921 to become one of the country's first airline pilots. In 1928, Smith and his crew in their Fokker trimotor *Southern Cross* made the first transpacific flight from the U.S. to Australia. Among a number of flight records, he made the first non-stop Australian transcontinental flight, the first flight between Australia and New Zealand, and the first eastward Pacific crossing from Australia to the U.S., in 1934 in the *Lady Southern Cross*. Smith and his co-pilot disappeared over the Andaman Sea in 1935.

4. "Heroic Woman Begins Flight to Australia in Light Aeroplane," *Sunday Times* [Sydney], 16 October 1927, 1.

5. Mrs. Keith Miller, "Only One Frock for Half a World's Journey," *The Mercury* [Hobart, Tasmania], 18 April 1928, 7.

6. Ibid.

7. Ibid.

8. Ibid.

9. Mrs. Keith Miller, "The Red Rose Flight: Tracing North Africa," *The Mercury*, 19 April 1928, 9.

10. Sirte would later gain another kind of recognition entirely as the birthplace of Muammar Gaddafi and the location of his death in 2011.

11. Miller, "The Red Rose Flight: Tracing North Africa," 9.

12. Ibid.

13. Ibid.

14. "To Australia: Mementoes of Flight," *The News*, 31 October 1927, 7.

15. Mrs. Keith Miller, "The Red Rose Flight: From Egypt to Baghdad," *The Mercury*, 20 April 1928, 5.

16. Ibid.

17. Ibid.

18. Ibid.

19. Ibid.

20. Mrs. Keith Miller, "Over Desert Wastes," *Morning Bulletin* [Rockhampton], 1 May 1928, 10.

21. Mrs. Keith Miller, "The Red Rose

Flight: Kaleidoscope of Mesopotamia," *The Mercury*, 24 April 1928, 5.

22. William R Polk, *The Elusive Peace: The Middle East in the Twentieth Century* (Abingdon, UK: Routledge, 2013), 58. Also, Warren Dockter, *Churchill and the Islamic World* (London: I. B. Tauris, 2015), 111–115.

23. Miller, "The Red Rose Flight: Kaleidoscope of Mesopotamia," 5.

24. Ibid.

25. Mrs. Keith Miller, "Flirting with Death: Risks on the Red Rose," *The Register*, 24 April 1928, 9.

26. Ibid.

27. Ibid.

28. Ibid.

29. His Majesty's Airships R100 and R101 never reached Karachi. The rigid airship R100, designed by a team headed by Barnes Wallis of bouncing bombs and "Dam Busters" fame, made a successful return crossing of the Atlantic to Canada in 1930. The R101, then the world's largest flying craft at 731 feet (223m), went down in France en route to India in October 1930. Of the fifty-four people on board, forty-eight died in the crash (compared to thirty-five in the later loss of the *Hindenburg*), including the Air Minister, Lord Thomson, government officials and most of the designers. Consequently, the R100 was broken up for scrap and the British airship program was over.

30. Miller, "Flirting with Death," 9.

31. King Amanullah Khan's attempts to modernize the country with the help of his wife Queen Soraya Tarzi, including granting equal access to education and rights for women, encountered conservative opposition. After he met Jessie and Bill, the King traveled through Europe meeting most of the monarchs then in power as well as Pope Pius XI. However, while he was absent rebellion erupted in the Kingdom, forcing the King to abdicate and flee to India and finally Switzerland.

32. "England to India: A Woman's Flight." *The Advertiser* [Adelaide], 8 December 1927, 15, and "Lady Aviator's Experiences," *Daily Mercury* [Mackay, Queensland], 9 December 1927, 4, are just two examples of many.

33. Miller, "Flirting with Death," 9.

34. Ralph Barker, *Verdict on a Lost Flyer: The Story of Bill Lancaster and Chubbie Miller* (Sydney: Fontana/Collins, 1986), 32. Also, Ralph Barker, "Mrs. Pugh," transcripts of interviews conducted c. 1968, Tape 3: 19.

35. Barker, "Mrs. Pugh," Tape 2: 3.

Chapter Three

1. Mrs. Keith Miller, "On Ross Smith's Trail: The Red Rose Over India," *the Register*, 25 April 1928, 12.

2. Maharajah Umaid Singh, an early proponent of flying and airplanes, had built the first airfield in Jodhpur in 1924. He would inaugurate the Jodhpur Flying Club in 1931 and within seven years there would be twenty-three airfields throughout the state. Today that flying club is one of the largest military air bases in Southeast Asia and his first airfield is an international airport. (See Peter Vacher, *The History of the Jodhpur Flying Club* [Apogee Prime, 2010].

3. Miller, "On Ross Smith's Trail," 12.

4. Ibid.

5. Ibid.

6. Ibid.

7. "Round the World Aviators: Plucky Mrs. Miller," *Newcastle Sun*, 20 December, 1927, 1.

8. Miller, "On Ross Smith's Trail," 12.

9. Ibid.

10. Ralph Barker, "Chubbie Pugh," transcripts of interviews conducted c. 1968, Tape 1: 19.

11. Mrs. Keith Miller, "The Red Rose: Damaged Engine and Not a Rupee," *The Mercury*, 27 April 1928, 5.

12. Said by local Myanmar legend to date back 2,500 years, the 325-foot-high Shwedagon Pagoda is the most sacred Buddhist pagoda in Myanmar, containing relics of four previous Buddhas. Enlarged over time, it was raised to its current height after an earthquake in 1768.

13. Barker, "Chubbie Pugh," Tape 1: 5.

14. Mrs. Keith Miller, "Our Flight to Australia: Damaged Engine and Cashless," *Table Talk*, 17 May 1928, 24.

15. Ibid.

16. Ibid.

17. The "Snake in the Cockpit" incident, as it became known, was an instant public relations success. Apart from Jessie's own account in Mrs. Keith Miller, "A Night of Misery: Red Rose on the Beach," *The Register* [Adelaide], 27 April, 1928, 10, some other catchy headlines are: "Snakes: Big Shock for Aviators," *Newcastle Sun*, 4

January 1928, 1; "Snake in Aeroplane," *Daily News* [Perth], 4 January 1928, 7; "Snake Goes Joy Riding," *Uralla Times* [New South Wales], 5 January 1928, 6.

18. "A Goodwill Flyer," *Daily Standard* [Brisbane], 6 January 1928, 7.

19. Miller, "A Night of Misery," 10.

20. Mrs. Keith Miller, "Red Rose Crashes," *The Register*, 28 April 1928, 11.

21. "The Red Rose Arrives," *Straits Times*, 7 January 1928, 9.

22. Captain W. N. Lancaster, "Hazardous: Globe Aviators' Most Critical Flight," *Newcastle Sun*, 9 January 1928, 1.

23. She was followed across the equator only a few weeks later by Lady Cobham, wife of Sir Alan Cobham, while they and their crew were flying between Entebbe on the northwest shore of Lake Victoria and Kisumu on the eastern shore in early February 1928. See Sir Alan Cobham's *Twenty Thousand Miles in a Flying Boat* (History Press, 2007).

24. Miller, "Red Rose Crashes," 11.

25. Muntok is now the site of the largest tin smelter in the world but is just as notorious as the site of the Muntok Internment Camp set up by the Japanese during World War II to house mainly British and Dutch civilians from the area. Lavinia Warner's film *Tenko* memorialized the women from the camp at Pelambang who formed a vocal orchestra, many of whom eventually died at Muntok and were buried there.

26. "Red Rose Crashes," *The News* [Adelaide], 10 January 1928, 1. Also,Miller, "Red Rose Crashes," 11.

27. Miller, "Red Rose Crashes," 11.

28. Barker, "Chubbie Pugh," Tape 1: 7.

29. Ralph Barker, *Verdict on a Lost Flyer: The Story of Bill Lancaster and Chubbie Miller* (Sydney: Fontana/Collins, 1986), 40.

30. Mrs. Keith Miller, "Our Flight to Australia," *Table Talk*, 24 May 1928, 27.

31. Mount Rinjani, or Gunung Rinjani, is a compound volcano that includes the regularly erupting Gunung Baru Jari. It is one of some 129 active Indonesian volcanoes and, along with three others, forms part of the Pacific Ring of Fire. As with most Indonesian volcanoes, the mountain is sacred.

32. Consequently, 1816 was known throughout Europe as the "Year without Summer," and as the stratospheric layer of volcanic sulphate aerosol cooled the Earth the Indian monsoon failed the following year as well as the rains in China, causing famine and cholera epidemics. Arctic sea ice and Alpine snows thawed in a nineteenth-century global warming. Read further, if you dare to challenge the myth that global climate change is a millennial phenomenon, in Gillen D'Arcy Wood's *Tambora* (Princeton: Princeton University Press, 2014).

33. Barker, "Mrs. Pugh," Tape 3: 19.

34. Mrs. Keith Miller, "Darwin at Last. Red Rose in the Rain," *The Register*, 30 April 1928, 12.

35. Ibid.

36. Ibid.

37. "Darwin Disappointed," *Northern Standard*, 20 March 1928, 2.

38. Miller, "Our Flight to Australia," *Table Talk*, 24 May 1928, 27.

Chapter Four

1. Ralph Barker, *Verdict On a Lost Flyer: The Story of Bill Lancaster and Chubbie Miller* (Sydney: Fontana/Collins, 1986), 41. Although Jessie stated quite clearly to Barker that it was a man who was the first to meet them in Australia, the local *Northern Standard*, 20 March 1928, 2, suggested it was the jail warden's wife, Mrs. Dempsey. However, on 23 March the newspaper apologized for getting that wrong. Apparently Mrs. W. M. Scott of Fannie Bay wanted to be recorded as the first person to have greeted them personally and to have obtained Jessie Miller's autograph, which she made sure included the line "First met by Mrs. Scott on arrival in Australia." Poor Mrs. Dempsey found herself quickly relegated to fourth place.

2. "Surprise Landing at Darwin Aerodrome," *Daily Standard* [Brisbane], 20 March 1928, 1.

3. This well-known saying, slightly altered here, now means "there's a problem." It is derived from Hamlet's "To be or not to be" soliloquy in Act III, Sc.1: "To die, to sleep, / To sleep, perchance to dream; / Aye, there's the rub." A rub was originally a dip or a bump in the surface of a lawn-bowls green.

4. "Popular Pair Inundated with Offers," *Brisbane Courier*, 21 March 1928, 17.

5. "Message from Mr. Bruce," *The News* [Adelaide], 20 March 1928, 1.

6. "Aviators Leave Darwin," *Northern Standard*, 23 March 1928, 4.

7. "Popular Pair Inundated with Offers," 17.
8. "Aviators Leave Darwin," 4.
9. *Brisbane Courier*, 26 March 1928, 13.
10. Hudson Fysh was not only the managing director but also the founder, with a group of investors in 1920, of the Queensland and Northern Territory Aerial Services Limited, now the airline better known by the acronym Qantas, made famous in more recent times in the film *Rain Man*. Originally based in the Queensland town of Winton, they moved to Longreach the following year. Connecting Longreach to Darwin, Qantas at that time maintained the longest direct airmail service in the world. (See John Gunn, *The Defeat of Distance: Qantas 1919-1939* (St Lucia, Queensland: University of Queensland Press, 1969), 27-28.
11. "Mrs. Miller: Praise from Sydney Women," *Chronicle* [Adelaide], 24 March 1928, 52.
12. "Country Women's Association," *Sunday Times* [Perth], 25 March 1928, 3S.
13. "Red Rose Arrives," *The Queenslander*, 29 March 1928, 44.
14. "Flight of the Red Rose," *The Telegraph* [Brisbane], 26 March 1928, 9.
15. "Lancaster and Mrs. Miller in Brisbane," *Brisbane Courier*, 27 March 1928, 13.
16. *Ibid.*, 13, 15.
17. "The Red Rose: Scene at Eagle Farm Aerodrome," *The Telegraph*, 27 March 1928, 9.
18. "Red Rose Survives Ordeal of Momentous Flight," *Daily Standard* [Brisbane], 27 March 1928, 1. Also, "In Brisbane: Warm Welcome," *Brisbane Courier*, 27 March 1928, 13, and, "The Red Rose," 9.
19. "Why No State Official Welcome to Mrs. Keith Miller?" *Truth*, 1 April 1928, 1.
20. "Red Rose Drops from the Blue," *Sunday Times*, 1 April, 1928, 1. Also, "Red Rose Fliers," *Sunday Mail* [Brisbane], 1 April 1928, 1.
21. "Red Rose: Arrival at Sydney," *Sydney Morning Herald*, 2 April, 1928, 11.
22. Barker, *Verdict*, 55.
23. "Mrs. Keith Miller: Mother in Sydney," *Daily Mercury* [Mackay, Queensland], 3 April 1928, 3.
24. "W.A. Airwoman: Mrs. Keith Miller," *The West Australian*, 3 April 1928, 5.
25. "Mrs. Keith Miller: Charming Personality," *Canberra Times*, 4 April 1928, 5.
26. "Red Rose Arrives," *The Argus* [Melbourne], 9 April 1928, 16.
27. *Ibid.*
28. *Ibid.*
29. *Ibid.*
30. Princess Anne of Lowenstein had been the second woman to disappear while on transoceanic flight, after Mildred Doran, when she, Captain Leslie Hamilton and Colonel Frederick Minchin vanished aboard the Fokker monoplane *St Raphael*, 31 August 1927, while attempting to be the first to fly the Atlantic from east to west. Elsie Mackay, also known as Poppy Wyndham, disappeared with Captain Hinchliffe off the coast of Ireland, 13 March, 1928, while attempting to fly across the Atlantic from west to east aboard the *Endeavour*.
31. "Red Rose Flyers Welcomed at Luncheon," *The Argus*, 12 April 1928, 6.
32. "Mrs. Keith Miller and Capt. Lancaster at the Majestic," *Table Talk*, 26 April 1928, 35. Child actress Louise Lovely was the first Australian female actor to be successful in America, where she made a number of films for Universal and Fox between 1915 and 1924. She returned to Australia to become involved in film production, but the industry fell into a slump and she and her second husband, Bert Cowan, eventually moved to Tasmania to manage the Prince of Wales Theatre in Hobart.
33. "Red Rose Due Tomorrow," *The Examiner* [Launceston], 27 April 1928, 7.
34. "Red Rose at Hobart: Enthusiastic Welcome by Great Crowd," *The Mercury*, 30 April 1928, 7.
35. Lancaster clearly stated when they arrived that it was Jessie who had piloted their aircraft across Bass Strait: "Red Rose: Arrival at Launceston," *The Age* [Melbourne], 30 April 1928, 11. Nevertheless, newspapers relentlessly pursued the point that this merely made her the fourth woman across the Strait in the air. Captain F. G. Huxley of Burnie claimed that his wife was the first woman carried across Bass Strait in an airplane, in January 1921. They had been followed a fortnight later by the mother and sister of Mr. Jack Pearce, all of whom were flying on a return trip to Melbourne with Lieutenant R. Parer (*Advocate* [Burnie], 24 April 1928, 4.)
36. "Pluck and Gentleness: Mrs. Keith Miller," *Examiner*, 30 April 1928, 9.
37. "The Miller Shorts," *The Western*

Champion [Barcaldine, Queensland], 5 May 1928, 12.

38. Prudence Black, "Fashion Takes Flight: Amy Johnson, Schiaparelli and Australian Modernism," *Hecate* 35 (2009): 62.

39. "The Red Rose," *The Mercury*, 1 May 1928, 7.

40. "The Red Rose: Flights Over Hobart," *The Mercury*, 3 May 1928, 7.

41. For more about Florence Taylor, see Chrystopher Spicer, *Great Australian World Firsts: The Things We Made, the Things We Did* (Sydney: Allen & Unwin, 2012), 241–251.

42. Ralph Barker, "Mrs. Pugh," transcripts of interviews conducted c. 1968, Tape 3: 10, 11.

43. In May 1928 *Red Rose* was reregistered G-AUTU, and a month later Lancaster sold the aircraft to R. A. Charlton of Sydney. In January 1929, the plane was sold to stockbroker J. R. Palmer of Sydney, who that month used the plane for one of the first "Flying Doctor" flights in Australia, taking a doctor from Mascot Aerodrome to Scone in New South Wales to perform a successful emergency operation. The plane was reregistered VH-UTU in August 1930. In January, 1931, it was involved in a bad landing accident near Lismore, NSW, and in April Leonard Palmer, who also used to fly the plane, was killed in the *Southern Cross Junior* at Mascot when a wing collapsed during an aerobatic loop. On 6 June 1936, piloted by John Horn, *Red Rose* crashed into a gully and burned when its engine failed while taxiing after landing near Singleton, NSW.

Chapter 5

1. "Woman Aviator," *The News* [Adelaide], 27 June 1928, 6.

2. "Lyon and Warner," *Queensland Times*, 9 July 1928, 7.

3. Ralph Barker, *Verdict on a Lost Flyer: The Story of Bill Lancaster and Chubbie Miller* (Sydney: Fontana/Collins, 1986), 47.

4. "Pacific Flyers Separate," *The West Australian*, 23 July, 1928, 13.

5. Back in San Francisco, a Mrs. Emma Meier Lyon was reported to have been granted a marriage annulment from Harry. She had produced a marriage certificate to one Harry Lyons, master mariner, who had listed his father as Henry and who had said his first wide had died. Emma had thought she was Harry's second wife until she discovered he was also married to Jane Lyon, with whom he was living in Maine. Emma's lawyer, Mr. J. McAtee, claimed that Harry had admitted to him that he had a wife still living and un-divorced. Harry replied he'd admitted no such thing and, furthermore, while he did know Emma he was happily married to Jane, couldn't understand what this was all about and presumed that Emma either had confused him with somebody else named Harry Lyons or was attempting to extort money from him. Finally, the Lyon family lawyer issued a statement that he'd subsequently received a letter from McAtee confirming Harry had never at any time been married to Emma (*Lewiston Evening Journal*, 18 August 1928, 6, and *San Jose News*, 17 August 1928, 13).

6. "Atlantic Flying Plans of Australians," *Barrier Miner* [Broken Hill], 4 October 1928, 1.

7. "New York to Bermuda," *The Age*, 30 October 1928, 9.

8. Ralph Barker, "Chubbie Pugh," transcripts of interviews conducted c. 1968, Tape 1: 11.

9. "Prohibition Law," *The Telegraph*, 21 December 1928, 20.

10. "Aviation Needs More Women," *Milwaukee Journal*, 18 November 1928, Section VI, 4.

11. *St. Petersburg Times*, 3 January 1929, and the *Glasgow Herald*, 11 January 1929. In April, Smith, at sixteen the youngest American woman at that time to be granted a pilot's license, would break that record by setting a new endurance record of 26.5 hours, and in November, with Bobby Trout, she stretched the endurance record even further to 42.5 hours.

12. "With an Avro 'Avian' ('Cirrus' Mk.III) in America," *Flight*, 14 March 1929, 215.

13. Red Bank Airport was a grass field established in 1926 by Air View Flying Service principally for aerial photography flights, but it quickly became a flight school and training field as well. Used into the 1970s for small airline and air taxi services, it did not survive the decade due to its proximity to housing.

14. "Mrs. Keith Miller Obtains Pilot's License in New York," *Longreach Leader*, 10 May 1929, 28.

15. *Matawan Journal* [New Jersey], 19 July 1929, 8. Also, Anthony Bianculli, *Iron*

Rails in the Garden State (Bloomington: Indiana University Press, 2008), 49.

16. Kathleen Winters, *Amelia Earhart: The Turbulent Life of an American Icon* (New York: Palgrave Macmillan, 2010), 92, for example, states Earhart held the fourth female commercial pilot's license in the U.S. after Ruth Nichols, Phoebe Omlie and Lady Heath and that at that time only forty women in the U.S. held any kind of pilot's license. For the *Liberty* list see Alicia Patterson, "I Want to Be a Transport," *Liberty*, 7 September 1929, 18–24. For the December figure, see *Women and Aviation*, no. 16, 29 December 1929.

17. "Across and Back: Atlantic Flight," *Newcastle Morning Herald and Miner's Advocate*, 11 April 1929, 5.

18. "Kingsford Smith for World Flight?" *Register News Pictorial* [Adelaide], 1 May 1929, 2.

19. "Kookaburra Compass," *The News*, 27 June 1929, 12. The deaths of Hitchcock and Anderson cast a shadow over the remainder of Kingsford Smith's aviation career. Although their bodies were eventually recovered from the desert, the wreck of the *Kookaburra* was lost until rediscovered by Dick Smith in 1978, after which it was moved to Alice Springs for display.

20. "The Southern Cross Inquiry," *The Mercury*, 25 June 1929, 7–8.

21. "Mrs. Keith Miller's Plans," *The Telegraph* [Brisbane], 8 August 1929, 6.

Chapter Six

1. Consolidated Aircraft would later design and manufacture the PBY Catalina and the B-24 Liberator bomber. A former World War I pilot, Fleet had established the first airmail service from Washington, D.C., to New York and then combined the assets of two aircraft companies to form Consolidated in 1923, primarily to manufacture military trainer aircraft. After his Fleet aircraft became such a success story, he sold that subsidiary back to Consolidated. By the 1930s, the Fleet was one of the most commonly used trainer and recreation aircraft in the U.S., usually priced at around $5,000. Lawrence 'Larry" Bell, probably more familiarly remembered as the founder of Bell Aircraft, had joined Consolidated after becoming managing director of the Glenn L. Martin Company. When Consolidated relocated to San Diego, he remained on the East Coast to form Bell Aircraft in 1935. The company became well-known for their fighter aircraft, helicopters and of course the Bell X-1, the first aircraft to break the sound barrier in level flight.

2. Mike Walker, *Powder Puff Derby: Petticoat Pilots and Flying Flappers* (Chichester: Wiley & Sons, 2003), 178.

3. Leslie E. Neville, "The Fleet Biplane," *Aviation*, 18 May 1929, 1686–90.

4. Walker, *Powder Puff Derby*, 178. Also, Ralph Barker, *Verdict on a Lost Flyer: The Story of Bill Lancaster and Chubbie Miller* (Sydney: Fontana/Collins, 1986), 54.

5. Gene Nora Jessen. *The Powder Puff Derby of 1929: The True Story of the First Women's Cross-Country Air Race* (Naperville, IL: Sourcebooks, ,2002), 119.

6. Walker, *Powder Puff Derby*, 178. Also, Ralph Barker, "Chubbie Pugh," transcripts of interviews conducted c. 1968, Tape 1: 12.

7. Louise McPhetridge Thaden, *High, Wide and Frightened* (Fayetteville: University of Arkansas Press, 2004), 46.

8. "Can't Trust Women," *Newcastle Sun*, 28 April 1927, 1.

9. *The Newcastle Sun* used the "Butterfly Sex" reference, 19 August, 1929, 1. Also, "Flight Around U.S.A.: 'She Derby' Continues," *Daily News* [Perth, Western Australia], 24 August 1929, 7.

10. Walker, *Powder Puff Derby*, 5.

11. *Ibid.*, 6, 8.

12. Jessen, *The Powder Puff Derby of 1929*, 59.

13. The crowd figures for the start of the Derby at Santa Monica vary wildly. Jessen, for example, could be said to be at the low end with a figure of 3,000 (*Ibid.*., 63), while Walker takes the high ground with a figure of 150,000 (*Powder Puff Derby*, 4). My figure is based on calculations by Heather Taylor, producer of the documentary *Breaking Through the Clouds*, who has researched the race for many years.

14. Thaden, *High, Wide and Frightened*, 46.

15. Jessen, *The Powder Puff Derby of 1929*, 66.

16. Opinion differs as to why von Mach was forced down. Jessen (*Ibid.*., 76) blames an overeager pilot in a stunt plane who buzzed her too closely. Taylor has pointed out to me other possible reasons: that von Mach's seat could have broken, that she got

lost, or that she feared for her safety at the overly crowded San Bernardino airport and turned around.

17. Thaden, *High, Wide and Frightened,* 48.

18. Walker, *Powder Puff Derby,* 178, and Jessen, *The Powder Puff Derby of 1929,* 119.

19. Jessen, *The Powder Puff Derby of 1929,* 119.

20. Ruth M. Reinhold, "The Old Douglas Airport," *Journal of Arizona History* 15, no. 4 (Winter 1974): 339. This story came directly from Walker, whom Reinhold interviewed in August 1972.

21. "Women's Air Derby: Hints at Sabotage," *The Mercury,* 23 August 1929, 9.

22. Thaden, *High, Wide and Frightened,* 53, and Amelia Earhart, *Last Flight* (New York: Crown, 1988), 8.

23. Because her body was found with her parachute still attached, the popular theory at the time in newspapers was that she had attempted to bail out of the plane for some reason and either the parachute failed to open or she did not have enough altitude for the parachute to be effective. However, Jessen (*The Powder Puff Derby of 1929,* 117, 128, 153) advocates the carbon monoxide poisoning theory.

24. As quoted in *Ibid..*, 129.

25. "Nerves on Edge," *The Examiner,* 23 August 1929, 4.

26. "Mrs. Thaden Now Leading Women's Big Aerial Race," *Schenectady Gazette* [New York], 21 August 1929, 1.

27. Jessen, *The Powder Puff Derby of 1929,* 156.

28. As related by Thaden, *High, Wide and Frightened,* 53–54.

29. "Caught in Whirlwind," *Schenectady Gazette,* 23 August 1929, 1.

30. Jessen, *The Powder Puff Derby of 1929,* 168.

31. "Wichita in Mighty Gesture of Welcome to Women Aviators" and, "Women Fliers Have Little to Say About 'Sabotage' Question," *Wichita Eagle,* 24 August 1929, 3.

32. Thaden, *High, Wide and Frightened,* 54.

33. Jessen, *The Powder Puff Derby of 1929,* 177–78.

34. "New Zealand Lady Derby Flyer Lands on Farm Near Xenia," *Evening Gazette* [Xenia], 26 August 1929.

35. Jessen, *The Powder Puff Derby of 1929,* 190.

36. *Dubbo Liberal and Macquarie Advocate,* 30 August 1929, 6.

37. Thomas G. Matowitz, *Cleveland's National Air Races* (Chicago: Arcadia, 2005), 17.

38. Margery Brown, "Why Women Pilots Have Organized," *Aeronautics,* March 1930, 35.

39. *Ibid.,* 36.

Chapter Seven

1. Ralph Barker, *Verdict on a Lost Flyer: The Story of Bill Lancaster and Chubbie Miller* (Sydney: Fontana/Collins, 1986), 55.

2. Prudence Black, "Fashion Takes Flight: Amy Johnson, Schiaparelli and Australian Modernism," *Hecate* 35 (2009): 35, 66.

3. Often known as "The Tin Goose," the all-metal Ford trimotor was developed by Ford in 1925, based on a design concept already introduced by Junkers in Germany. In fact, Ford's airplane was so similar that Junkers sued them for patent infringement and won, not just once but twice! Like Junkers' plane, Ford's American version was a success because it could be readily converted from passenger to freight to military service, to land on water, snow or land, and it had a reputation for durability in all weather conditions.

4. These details of the scoring system are from John H. Livingston. *One-Two: The Story of the Fifth National Air Tour as Related by the Winner* (Troy, OH: Waco Aircraft Company, 1930; St. Paul, MN: Aviation Foundation of America, 2002), 5–7.

5. Lesley Forden, *The Ford Air Tours 1925-31* (New Brighton, MN: Aviation Foundation of America, 2003), 87, 110–12.

6. *Ibid.,* 88.

7. A Ford Air Tour veteran, Frank Hawks was a military pilot before becoming a barnstormer and instructor who famously took a young Amelia Earhart on her first flight. A tireless self-promoter, he was eventually hired by the Texas Company (Texaco) in 1927 to head their marketing program and in the following years proceeded to compete in races and to establish a number of flying records in the U.S. and Europe, including setting the transcontinental speed record in 1929, which he broke twice in his own lifetime.

8. "Reliability Air Tour Planes Due

Here Tomorrow," *Montreal Gazette*, 5 October 1929, 19.

9. "Noted Aviatrix Flies in Derby," *Calgary Daily Herald*, 17 October 1929, 23.

10. Livingston began his lifetime career in aviation as an airplane mechanic but was soon flying as a barnstormer. By 1922, he became Field Operations Manager for Mid-West Airways in Illinois. Six years later, he bought the company. His first Air Tour was in 1926 and within seven years he reputedly won more races than any other pilot: during his life he placed first in eighty national air races.

11. "Montreal's First Air Pageant Drew 40,000 Spectators," *Montreal Gazette*, 7 October 1929, 14. Also, Livingston, *One-Two*, 10–11.

12. "Reliability Fliers Reach Springfield," *Montreal Gazette*, 8 October 1929, 1. Also, Livingston, *One-Two*, 12.

13. Livingston, *One-Two*, 13.

14. "Mainly About People," *Daily News* [Perth], 21 November 1929, 11.

15. Forden, *The Ford Air Tours*, 89.

16. Livingston, *One-Two*, 15.

17. *Ibid.*, 21.

18. "Livingston Wins National Air Tour," *Lewiston Daily Sun* [Lewiston, Maine], 22 October 1929, 1.

19. Ralph Barker, "Chubbie Pugh," Transcripts of Interviews Conducted C. 1968, Tape 1: 14.

20. "First Victor Plane Built," *Brooklyn Daily Eagle*, 15 December 1929, 46. Also, "Mrs. Keith Miller," *The Worker* [Brisbane], 8 January 1930, 19. Heinrich and his brother Arthur were speedboat designers before turning to aircraft design, inspired by the Wright brothers. In May 1910, Heinrich made the first flight in their co-designed monoplane, and the brothers went on to set up the Victor Aircraft Corporation as well as two flying schools.

21. Barker, "Mrs. Pugh," Tape 3: 24.

22. "Aviation News: Join Sales Staff," *Brooklyn Daily Eagle*, 6 February 1930, 11.

23. Barker, *Verdict*, 57–58.

24. One of the first female airmail pilots in the U.S. and Canada, Katherine Stinson held distance and endurance records in both countries during her career. Eddie's Stinson Aircraft Company later became the Stinson Aircraft Corporation based in Detroit, manufacturers of the very successful Detroiter and Stinson aircraft.

25. Barker, *Verdict*, 58.

26. Barker, "Chubbie Pugh," Tape 1: 18.

27. "Mrs. Keith Miller Spends Night on Rolling Wave," *Evening News* [Sydney], 29 May 1930, 1.

28. Barker, "Chubbie Pugh," Tape 1: 18.

29. Johnson also became the first Englishwoman to obtain a ground engineer's license. In 1932 she married pilot Jim Mollison and so she was also known as Amy Mollison. In 1932 she set a solo record for the flight from London to Cape Town and then in 1933 with her husband flew from Wales to the U.S. In 1934 they set a record from England to India. In 1940 she joined the newly formed Air Transport Auxiliary ferrying RAF aircraft, only to be lost at sea when her airplane went down in the Thames Estuary in 1941.

30. "The Week," *Adelaide Chronicle*, 3 July 1930, 42.

31. "Australian Airwoman to Fly Atlantic Alone This Month," *The Observer* [Adelaide], 19 June 1930, 15.

32. Barker, *Verdict*, 59.

Chapter Eight

1. Ralph N. Miller, "Remember the Bullet?" *Popular Aviation*, January 1939, 33, 76–78.

2. Amelia Earhart, *Last Flight* (New York: Crown, 1988), 11.

3. *The Free-Lance Star* [Fredericksburg, Virginia], 29 November 1930, 1.

4. Ralph Barker, "Chubbie Pugh," transcripts of interviews conducted c. 1968, Tape 1: 15.

5. *Ibid.*, Tape 3: 24.

6. "On Their Dignity," *Newcastle Sun*, 19 August 1930, 3.

7. "Mrs. Keith Miller: New York to Los Angeles," *The Telegraph* [Brisbane], 14 October 1930, 5.

8. "Coast to Coast," *Brisbane Courier*, 18 October 1930, 13.

9. "Mrs. Miller Modest," *The News*, 28 October 1930, 7.

10. Barker, "Chubbie Pugh," Tape 1: 15.

11. "Bent Propeller: Mrs. Miller 'Playing Safe,'" *The Telegraph*, 25 October 1930, 3.

12. "Mrs. Keith Miller: New Women's Record," *The Telegraph*, 27 October 1930, 5.

13. *Pittsburgh Press*, 2 November 1930, 1, 4.

14. E. M. Fitzgerald, "Pouf! Who Says 13

Is Bad Luck?" *Pittsburgh Press*, 12 November 1930, 1–2.

15. *The News and Courier* [Charleston, South Carolina], 29 November 1930, 2.

16. Ibid.

17. Ibid.

18. *Pittsburgh Press*, 24 November 1930, 2. The fifth Cuban president, Machado had been a general during the 1895–98 Cuban War of Independence. A public works and infrastructure advocate, he was responsible for the island's Central Highway and Capitol Building, but he, too, fell under the spell of power, eventually closing the university and discouraging free speech. It was during his term that the first Cuban links to Stalinist communism were formed. He stepped down in the face of an army revolt in 1933, to die an exile in Miami seven years later.

19. Homer E. Knoblaugh, "Missing Woman Flier Had Premonition of Tragedy," *The Free Lance-Star* [Fredericksburg, Virginia], 29 November 1930. 1.

20. "Hopes Fade in Search for Mrs. Miller," *Newcastle Sun*, 1 December 1930, 8. Ingalls went on to win the Harmon Trophy as the most outstanding aviatrix of 1934. She made the first flight over the Andes by an American woman, the first solo flight by a woman from North to South America and set a new solo distance record for an aviatrix of some 17,000 miles. Sadly, her illustrious flying career did not survive a conviction for espionage during World War II, guilt for which Ingalls always denied, claiming she was naïvely attempting to work undercover to gather information about German activities in America. It's quite likely that her vocal opposition to America entering the war had not earned her any friends in high places. Applications for a post-war pardon were consistently refused.

21. "Woman Aviator Missing," *The Mail* [Adelaide], 29 November 1930, 1.

22. Mrs. Jessie Keith Miller, "Mrs. Keith-Miller Tells Own Story of Air Adventure," *Pittsburgh Post-Gazette*, 3 December 1930, 1. See also Ralph Barker, *Verdict on a Lost Flyer* (Sydney: Fontana/Collins, 1986), 62–64.

23. Miller, "Mrs. Keith-Miller Tells Own Story of Air Adventure," 4.

24. Ibid.

25. M. P. "Storky" Adams, Australia's Old Man of the Sea, *The Sporting Globe* [Melbourne], 26 April 1941, 6. The ruins of Cavill's house on Bimini, known as Conch House, can still be visited along the Bimini Nature Trail on South Bimini.

26. Jessie's account of her meeting with the Forsythes and Cavill is from Barker, "Chubbie Pugh," Tape 1, 20–21.

27. "Mrs. Keith-Miller Is Safe: Lands on Island in Storm," *Pittsburgh Post-Gazette*, 2 December 1930, 1.

28. "Husband's Relief," *Newcastle Sun*, 2 December 1930, 1.

29. "Mrs. Keith-Miller Is Safe," 4.

30. "Report Flier in Publicity Stunt Denied," *Pittsburgh Post-Gazette*, 4 December 1930, 17.

31. "Woman Aviator Found Safe on Bahama Island," *Telegraph-Herald and Times-Journal* [Dubuque, Iowa], 2 December 1930, 1.

32. "Seeks American Citizenship," *Cairns Post*, 3 December 1930, 4.

33. "Mrs. Keith Miller: Unhappy and Ashamed," *The Telegraph* [Brisbane], 4 December 1930, 8.

34. "Aviatrix Plans to Finish Hop," *Pittsburgh Press*, 4 December 1930, 2.

35. *Telegraph Herald and Times Journal*, 5 December 1930, 8.

36. "Woman Aviator Lands in Miami," *St. Petersburg Times*, 6 December 1930, 8.

37. "Mrs. Keith Miller Plane Wrecked in Crash," *Sydney Morning Herald*, 16 December 1930, 11.

38. Ralph Barker, *Verdict on a Lost Flyer: The Story of Bill Lancaster and Chubbie Miller* (Sydney: Fontana/Collins, 1986), 66.

39. *Border Cities Star*, 19 December 1930, 8.

40. "Mrs. Keith-Miller to Seek Medical Aid in New York," *Pittsburgh Post-Gazette*, 19 December 1930, 17.

41. "Competition Letters," *Freeman's Journal* [Sydney], 19 February 1931, 43.

Chapter Nine

1. "Will Become an American," *The News*, 9 June 1931, 7.

2. "Taken as Favor," *The Mail* [Adelaide], 27 June 1931, 2.

3. Ralph Barker, "Mrs. Pugh," transcripts of interviews conducted c. 1968, Tape 3: 15.

4. There would be another Mrs. Keith Miller, as Keith Miller married again in 1938 to Tasmanian solicitor Nancy Grant, daughter of State Senator C. W. Grant, and they

settled on Findon Avenue, Caulfield, in Melbourne. An educated woman in her own right with two degrees and widely traveled, including to Soviet Russia, Grant had established the first female legal partnership in Victoria two years previously. However, Keith died suddenly in 1943 while in Hobart. They did not, apparently, have any children ("Obituary: Mr. Keith Miller," *The Age*, 25 March 1943, 3).

5. "My Career Means Too Much to Me," *New Call* [Perth], 19 May 1932, 3.

6. "Big Hotel Fire," *Daily News* [Perth], 31 July 1931, 7. Also, *Pittsburgh Post-Gazette*, 31 July 1931, 2, and "Fort Plain Hotel Ruined by Flames, May Not Be Rebuilt," *Utica Daily Press* [NY], 31 July 1931, 15.

7. "Never Been in Air," *Maryborough Chronicle*, 13 October 1931, 5.

8. Barker, "Chubbie Pugh," Tape 1: 22.

9. Mrs. Keith Miller, "Mrs. Keith Miller Continues Her Story," *Daily Express* [London], 28 October 1932, 3.

10. *Ibid.*
11. *Ibid.*
12. Barker, "Mrs. Pugh," Tape 2: 1.

13. "Lancaster Pleased at Meeting with Clarke," *Rochester Evening Journal*, 5 May 1932, 15. Bill's diary entries were published in newspapers across the U.S. and in Australia as a result of his later arrest and consequent seizure of his documents. Just how those documents came to fall into the hands of newspaper reporters was never specifically clarified.

14. *Ibid.*
15. *Ibid.*
16. Barker, "Mrs. Pugh," Tape 2: 3.
17. *Ibid.*, Tape 3: 5.

18. Eudora Ramsay Richardson, "The Ladies of the Lobby," *North American Review* 227, no. 6 (June 1929): 648–55. Her obituary appeared as: "Mrs. T.H. Clarke, Writer on Feminism," *New York Times*, 29 August 1956, 28.

19. Miller, "Mrs. Keith Miller Continues Her Story," 3; Barker, "Mrs. Pugh," Tape 2: 1.

20. Barker, "Mrs. Pugh," Tape 3: 3, and "Lancaster Pleased at Meeting with Clarke," 15.

21. Barker, "Mrs. Pugh," Tape 3: 5.
22. *Ibid.*, 15.

23. Ralph Barker, *Verdict on a Lost Flyer: The Story of Bill Lancaster and Chubbie Miller* (Sydney: Fontana/Collins, 1986), 68.

24. Barker, "Mrs. Pugh," Tape 2: 3.
25. *Ibid.*, Tape 3, 4.

26. *The Mail* [Adelaide], 6 August 1932, 3. These entries were read in court as evidence.

27. Barker, "Mrs. Pugh," Tape 3: 6.
28. Barker, *Verdict*, 91–92.
29. Barker, "Chubbie Pugh," Tape 3, 22–23.
30. Miller, "Mrs. Keith Miller Continues Her Story," 3.
31. Barker, "Mrs. Pugh," Tape 3: 15.
32. *Ibid.*, 4.
33. Barker, *Verdict*, 84.

34. "Did Not Smuggle Chinese," *Daily Standard* [Brisbane], 15 August 1932, 1. Also, "Flyer's Diary Reveals Love for Aviatrix, *Rochester Evening Journal*, 4 May 1932, 11.

35. *Ibid.*

36. Rex Saffer, "Threats Made Against Youth, Witness Says," *St. Petersburg Times*, 5 August 1932, 5.

37. *Milwaukee Sentinel*, 4 May 1932, 1.
38. *Ibid.*

39. Rex Saffer, "Flyer's Diary Given to Jury During Trial," *St. Petersburg Times*, 6 August 1932, 1.

40. *Ibid.* and *The Mail* [Adelaide], 6 August 1932, 3.

41. *Ibid.*
42. *Milwaukee Sentinel*, 4 May 1932, 1.

43. Rex Saffer, "Tangled Story of Loves Told by Girl Flyer," *St. Petersburg Times*, 4 August 1932, 1.

44. This conversation between Jessie, Bill and Haden was part of her court testimony recorded in the *Pittsburgh Press*, 4 August 1932, 8.

45. Saffer, "Threats Made Against Youth," 5.

46. "Flier Tells of Suicide Pact," *Pittsburgh Press*, 4 August 1932, 1.

Chapter Ten

1. The following account of the period leading up to Haden's death is drawn from Jessie's own recounting in her article "Mrs. Keith Miller Tells of Her Police Ordeal," *Daily Express*, 31 October 1932, 3, as well as testimony by Jessie, Huston and other witnesses at Lancaster's subsequent trial reported in "Mrs. Keith Miller Bares Love Triangle at Trial," *Pittsburgh Press*, 3 August 1932, 1, 8.

2. Miller, "Mrs. Keith Miller Tells of Her Police Ordeal," 3.

3. Mrs. Keith Miller, "I Appeal to You!" *Daily Express*, 27 October 1932, 10.

4. This account of her detention and interrogation is also drawn from Jessie's *Daily Express* articles of 3 August, 1, and 27 October, 10.

5. "I Loved Him, I Was Alone," *Newcastle Sun*, 23 April 1932, 1.

6. "Airman's Death," *Sunday Times* [Perth], 24 April 1932, 1S.

7. "Fiance of Aviatrix Slain, Doctors Say," *Pittsburgh Press*, 23 April 1932, 2.

8. Ralph Barker, *Verdict on a Lost Flyer: The Story of Bill Lancaster and Chubbie Miller* (Sydney: Fontana/Collins, 1986), 112.

9. Miller, "Mrs. Keith Miller Tells of Her Police Ordeal," 3.

10. "Mother Hears Aviator Deny He Killed Son," *Pittsburgh Press*, 25 April 1932, 5.

11. "Everything Has Now Changed," *Daily News* [Perth], 27 April 1932, 7.

12. Barker, *Verdict*, 114.

13. Ralph Barker, "Mrs. Pugh," transcripts of interviews conducted c. 1968, Tape 2: 5.

14. Barker, *Verdict*, 114.

15. Barker, "Mrs. Pugh," Tape 2: 5.

16. "Lancaster Held as Miami Murderer," *New York Times*, 3 May 1932, 1.

17. Ibid.

18. Mrs. Keith Miller, "Amazing War Service Ban on Murder Jury," *Daily Express*, 1 November 1932, 3.

19. Barker, "Mrs. Pugh," Tape 2: 6.

20. Miller, "I Appeal to You!" 10.

21. "Mother Thinks Haden Clarke Was Murdered," *St. Petersburg Times*, 8 May 1932, Sec. 1, 6.

22. Miller, "Amazing War Service Ban on Murder Jury," 3.

23. "Mrs. Miller Arrested," *Daily News* [Perth], 24 June 1932, 7.

24. "Australian Airwoman Speaks Out Before Murder Trial," *Daily News*, 1 August 1932, 7.

25. Miller, "Amazing War Service Ban on Murder Jury," 3.

26. Ibid. and *Palm Beach Post*, 3 August 1932, 1.

27. The following details of and quotations from Carson's opening speech in court are taken from his transcript published in James M. Carson, *The Lancaster Case: Arguments for the Defense* (Miami, 1934), 9–35.

28. Barker, "Mrs. Pugh," Tape 2: 8.

29. "Mrs. Keith-Miller Tells of Two Loves," *New York Times*, 4 August 1932, 14. Also, "Mrs. Miller Tells of 'Fool Suicide Pact,' " *Daily Express*, 5 August 1932, 7.

30. "Love, Jealousy and Tragedy," *Daily Standard*, 5 August 1932, 1.

31. "Flyer's Love Notes Given Jury," *Pittsburgh Press*, 5 August 1932, 1.

32. Larry Smith, "Killing Story Retold by Girl Flier," *Rochester Evening Journal*, 4 August 1932, 1.

33. "Lancaster's Grief at Jilting Shown," *New York Times*, 6 August 1932, 26.

34. Rex Saffer, "Threats Made by Lancaster," *St. Petersburg Times*, 7 August 1932, 3.

35. Barker, "Mrs. Pugh," Tape 2: 8.

36. Jessie's Monday morning testimony is from accounts in "They Are Crucifying Me," *Daily Standard*, 9 August, 1932, 1. Also, "State Rests Case Against Lancaster," *Lewiston Daily Sun*, 9 August 1932, 1, 4. Also, "Mrs. Keith Miller Loved Clarke Until He Lied to Her," *Grape Belt and Chautauqua Farmer* [Dunkirk, New York], 9 August 1932, 4., and, "Mrs. Keith Miller Is Disillusioned," *Ottawa Citizen*, 8 August 1932, 2.

37. Barker, *Verdict*, 150.

38. "Lancaster's Story of His Love for Mrs. Miller," *Daily Standard*, 10 August 1932, 1.

39. "Lancaster Murder Trial," *Daily Standard*, 10 August 1932, 6. Also, "Slain Man Warned of Aviatrix' Weaknesses," *Rochester Evening Journal*, 9 August 1932, 49.

40. "Lancaster's Story of His Love for Mrs. Miller," 1.

41. "Didn't Kill Clarke," *The News*, 11 August 1932, 1,9.

42. "Trial of Lancaster," *Sydney Morning Herald*, 12 August 1932, 10, and, "Lancaster Tells of Clarke's Death," *New York Times*, 11 August, 1932, 6. Also, *Ottawa Citizen*, 10 August 1932, 13.

43. "Love Notes to Clarke," *The News*, 12 August 1932, 1.

44. "Clarke Is Pictured as Narcotic Addict," *New York Times*, 12 August 1932, 13.

45. "Love Notes to Clarke," 1.

46. "Clarke Is Pictured as Narcotic Addict," 13.

47. Colin Evans. *A Question of Evidence* (New York: Barnes & Noble Books, 2003), 60. Most of this section referring to Hamilton's evidence, unless otherwise indicated, is sourced from the fifth chapter, "William

Lancaster (1932): A Bullet in the Night," a case to which Evans refers as perhaps Hamilton's greatest "success."
48. Rex Saffer, "Murder Jury Shown Skull of Dead Man," *St. Petersburg Times*, 13 August 1932, 8.
49. *Ibid.*
50. "Lancaster Denies Indicating Guilt," *New York Times*, 14 August 1932, 13.
51. "Closing Argument in Lancaster Trial," *New York Times*, 16 August, 18. Also, *San Jose News*, 15 August 1932, 2.
52. This entire closing speech was published in Carson's *The Lancaster Case*, 36–118, from which the following section about it has been sourced. All quotations refer to that version unless otherwise noted.
53. Observed by *New York Times* journalist in "Makes 4-Hour Plea to Save Lancaster," 17 August 1932, 18.
54. Mrs. Keith Miller, "Storm of Joy at Airman's Acquittal," *Daily Express*, 2 November 1932, 3.
55. "Lancaster Acquitted of Murder," *Daily Standard*, 18 August,1932, 1.
56. "Copyrighted Comment," *Florence Times* [Florence, AL], 25 August 1932, 2.
57. "Lancaster Is Freed," *New York Times*, 18 August 1932, 17.
58. *Ibid.*
59. Larry Smits, "Captain Lancaster, Acquitted, to Seek Job," *Rochester Evening Journal and the Post Express*, 18 August 1932, 4.
60. "I Knew Old Bill Would Come Through," *Daily Standard*, 19 August 1932, 5.
61. *Sunday Times* [Perth], 4 March 1934, 6.
62. Prudence Black, "Fashion Takes Flight: Amy Johnson, Schiaparelli and Australian Modernism," *Hecate* 35 (2009): 70.
63. *Ibid.*, 71.

Chapter Eleven

1. "Case Against Mrs. Miller," *Daily Standard*, 20 August 1932, 1.
2. "Queens of the Air Exhausted by Long Flight," *Pittsburgh Press*, 23 August 1932, 15.
3. "Conspiracy Charge or Deportation?" *The News*, 3 September 1932, 6.
4. "Will Marry Her Whether She Wants to or Not," *The News*, 21 September 1932, 8.
5. "Longer in U.S.A.," *Queensland Times*, 17 September 1932, 6.

6. "Will I Ever Live It Down?" *The Mail*, 15 October 1932, 1.
7. "Mrs. Keith Miller and Flier Off for England Soon," *The Mail*, 8 October 1932, 1.
8. "Will I Ever Live It Down?" 1.
9. "Not Deported," *Saturday Night Telegraph* [Brisbane], 15 October 1932, 11.
10. Ralph Barker, "Mrs. Pugh," transcripts of interviews conducted c. 1968, Tape 2: 9.
11. Mrs. Keith Miller, "I Appeal to You!" *Daily Express*, 27 October 1932, 10.
12. *Ibid.*
13. Mrs. Keith Miller, "Mrs. Keith Miller Continues Her Story," *Daily Express*, 28 October 1932, 3.
14. Mrs. Keith Miller, "Court Storm of Joy at Airman's Acquittal," *Daily Express*, 2 November 1932, 3.
15. Barker, "Mrs. Pugh," Tape 3: 11.
16. *Ibid.*
17. Ralph Barker, *Verdict on a Lost Flyer: The Story of Bill Lancaster and Chubbie Miller* (Sydney: Fontana/Collins, 1986), 208.
18. Barker, "Mrs. Pugh," Tape 2: 9.
19. *The Advertiser*, 12 April 1933, 19.
20. Barker, "Mrs. Pugh," Tape 3: 11.
21. *Ibid.*
22. "Capt. Lancaster Leaves for Cape," *The Advertiser* [Adelaide], 12 April 1933, 19.
23. "Mother's Farewell to Her Son," *Daily Express*, 20 May 1933, 1.
24. "The Diary of Bill Lancaster: An Incredible Story of Courage," *Daily Express*, 22 October 1962, 9.
25. Barker, *Verdict*, 209–211.
26. Barker, "Mrs. Pugh," Tape 2: 9.
27. Barker, *Verdict*, 213.
28. *Ibid.*, 214.
29. *Barrier Miner*, 19 April 1932, 1.
30. "Captain Lancaster Search Commenced," *Sydney Morning Herald*, 18 April 1933, 6.
31. "Mystery Airman. Is It Captain Lancaster?" *Central Queensland Herald*, 4 January 1934, 50.
32. "Lancaster's Mysterious Fate," *The World's News* [Sydney], 5 July 1933, 4.
33. *Courier Mail* [Brisbane], 5 December 1933, 10.
34. "Flyers Who Disappear," *The Mercury* [Hobart], 24 February 1936, 8.
35. Barker, "Mrs. Pugh," Tape 3: 9.
36. "Reflection on Airman's Widow," *The Telegraph* [Brisbane], 6 April 1934, 2.
37. Known popularly as the "Greta

Garbo of the Skies" and honored by the Maori people with the title "Daughter of the Skies," Batten was inspired to fly at eighteen when Kingsford Smith took her up in the *Southern Cross*. Like Johnson and Jessie, she'd only had her license a short time before attempting flights to Australia. In 1936 she became the first woman to fly from England to New Zealand, but although she received a number of international awards, like Jessie her career didn't survive World War II, and she died a recluse on the island of Majorca.

38. "Cannot Get Backing," *The News*, 7 June 1934, 9.
39. "Ambitious Airwomen," *Sydney Morning Herald*, 2 Thursday 1934, 12S.
40. "World's Greatest Air Race," *The Mail* [Adelaide],1 September 1934, 1S.
41. "Mrs. Miller Seeks Funds," *Courier Mail* [Brisbane], 22 September 1934, 15.
42. *Northern Standard* [Darwin], 9 October 1934, 2.
43. Barker, "Mrs. Pugh," Tape 2: 12.
44. "Women in World Air Race," *Australian Women's Weekly*, 24 March 1934, 13.
45. Ibid.
46. Ibid.
47. "Mrs. Bruce's Bad Luck," *Daily News* [Perth], 12 August 1932, 7.
48. Barker, "Mrs. Pugh," Tape 2: 13.
49. "1932 Birthday Honours," *Flight*, 10 June 1932, 515.
50. "Pilot Saves Women in Blazing Plane," *Daily Mirror* [London], 15 July 1933, 3.
51. Barker, "Mrs. Pugh," Tape 2: 13.
52. Ibid., 10.
53. "The Robinson Redwing," *Motor Sport*, September 1930, 38–40.
54. Jessie wrote about the first part of her trip up to her arrival in Bechar in "A Woman in the Air," *Australian Woman's Mirror*, 12 November 1935, 13.
55. Ibid., 49.
56. Jessie wrote about the second part of her trip, from Bechar to Cotonou, in "Flying Over the Desert Sands of Africa," *Australian Woman's Mirror*, 26 November 1935, 13, 23.
57. This account of the visit to Dahomey and the dance appeared as Mrs. Keith Miller, "Night-Piece in Dahomey," *Australian Woman's Mirror*, 17 December 1935, 13.
58. Ibid., 13.
59. "Noted Woman Flyer Is Now a Typiste," *Truth* [Brisbane], 5 January 1936, 19.
60. John Pugh, "Five Miles Up Hunting the Weather," *Sunday Mail* [Brisbane], 25 August 1935, 32.
61. "Airmen Rescued in Channel," *Yorkshire Post* [UK], 13 May 1935, 10.
62. "Mrs. Keith Miller to Marry Again," *Townsville Daily Bulletin*, 4 April 1936, 8.
63. "Women's Flying Corps: Australian's Scheme," *Townsville Daily Bulletin*, 30 May 1936, 7.
64. Ibid.
65. "Women Better Pilots than Men," *Daily News*, 24 June 1936, 3.
66. Nancy Bird Walton, *My God! It's a Woman* (Sydney: Harper Collins, 2002), 117–118.
67. *Supplement to the London Gazette*, 1 January 1944, 8.
68. *Supplement to the London Gazette*, 9 June 1955, 3292.

Epilogue

1. This conversation is as remembered by Jessie in Ralph Barker, "Mrs. Pugh," transcripts of interviews conducted c. 1968, Tape 3: 1, and referred to in Ralph Barker, *Verdict On a Lost Flyer: The Story of Bill Lancaster and Chubbie Miller* (Sydney: Fontana/Collins, 1986), 218.
2. Barker, *Verdict*, 9–11.
3. Barker, "Mrs. Pugh," Tape 3: 1.
4. Barker, *Verdict*, 219.
5. Barker, "Mrs. Pugh," Tape 3: 1.
6. "The Diary of Bill Lancaster: An Incredible Story of Courage," *Daily Express*, 22 October 1962, 9.
7. "The Last Diary of Bill Lancaster," *Daily Express*, 23 October, 1962, 16.
8. "The Last Diary of Bill Lancaster," *Daily Express*, 24 October, 1962, 6.
9. "The Last Diary of Bill Lancaster," *Daily Express*, 25 October, 1962, 10.
10. "The Last Diary of Bill Lancaster," *Daily Express*, 26 October, 1962, 8.
11. Barker, "Mrs. Pugh," Tape 2: 11.

Bibliography

Articles by Jessie Miller (chronological)

The Pacific

"A Pacific Paradise: Some Impressions of Noumea." *Table Talk* [Melbourne], 13 May 1926, 16–17.

England to Australia

Miller's account of the England to Australia flight was serialized in a number of newspapers, though the articles were often cut to fit available space. As the location of Miller's original manuscript is unknown at the time of writing this book, I compared as many versions as possible in order to collate a definitive set of installments that provide the most intact and inclusive source of information about this journey.

"Only One Frock for Half a World's Journey." *The Mercury* [Hobart], 18 April 1928, 7. (I)
"The Red Rose Flight: Tracing North Africa." *The Mercury*, 19 April 1928, 9. (II)
"The Red Rose Flight: From Egypt to Baghdad." *The Mercury*, 20 April 1928, 5. (III)
"Over Desert Wastes." *Morning Bulletin* [Rockhampton, Queensland], 1 May 1928, 10. (III)
"The Red Rose Flight: Kaleidoscope of Mesopotamia." *The Mercury*, 24 April 1928, 5. (IV)
"Flirting with Death: Risks on the Red Rose." *The Register* [Adelaide], 24 April 1928, 9. (V)
"On Ross Smith's Trail: The Red Rose Over India." *The Register*, 25 April 1928, 12. (VI)
"The Red Rose: Damaged Engine and Not a Rupee." *The Mercury*, 27 April 1928, 5. (VII)
"A Night of Misery: Red Rose on the Beach." *The Register*, 27 April 1928, 10. (VIII)
"Red Rose Flight. Snake as Passenger in Cockpit: New Use for Joystick." *The Mercury*, 30 April 1928, 6. (VIII)
"Red Rose Crashes." *The Register*, 28 April 1928, 11. (IX)
"Darwin at Last. Red Rose in the Rain." *The Register*, 30 April 1928, 12. (X)

England to Australia in *Table Talk* Magazine

This magazine published an illustrated version of the preceding series, with maps, compressed into the following five articles.

"Our Flight to Australia." *Table Talk*, 19 April 1928, 8–9.
"Our Flight to Australia: The First White Woman in the Desert." *Table Talk*, 26 April 1928, 13–16.
"Our Flight to Australia: In the Garden of Eden." *Table Talk*, 3 May 1928, 13–14.
"Our Flight to Australia: Damaged Engine and Cashless." *Table Talk*, 17 May 1928, 24–26.
"Our Flight to Australia: The End of an Eventful Flight." *Table Talk*, 24 May 1928, 24–27.

Flight to Havana

"Mrs. Keith-Miller Tells Own Story of Air Adventure." *Pittsburgh Post-Gazette*, 3 December, 1, 4.

253

The Lancaster Trial

"I Appeal to You!" *Daily Express*, 27 October 1932, 10.
"Mrs Miller Continues Her Story." *Daily Express*, 28 October 1932, 3.
"Mrs Keith Miller Tells of Her Police Ordeal." *Daily Express*, 31 October 1932, 3.
"Amazing War Service Ban on Murder Jury." *Daily Express*, 1 November 1932, 3.
"Court Storm of Joy at Airman's Acquittal." *Daily Express*, 2 November 1932, 3.

Flight to Africa

"A Woman in the Air." *Australian Woman's Mirror*, 12 November 1935, 13, 49.
"Flying Over the Desert Sands of Africa." *Australian Woman's Mirror*, 26 November 1935, 13, 23.
"Night-Piece in Dahomey." *Australian Woman's Mirror*, 17 December 1935, 13, 36.

Books

Barker, Ralph. *Verdict on a Lost Flyer: The Story of Bill Lancaster and Chubbie Miller*. Sydney: Fontana/Collins, 1986.
Beveridge, Sidney A. *The Story of the Beveridge Families of England and Scotland*. Melbourne, Victoria: McLaren, 1923.
Bianculli, Anthony. *Iron Rails in the Garden State*. Bloomington: Indiana University Press, 2008.
Burgess, Trish, ed. *Bearing the Heart of a Sailor: Letters from the Antarctic and Other Faraway Places from Tom Oates to Elizabeth Eadie—1936-1938*. Fisher, ACT: 2004.
Copley, Greg. *Australians in the Air*. Adelaide: Rigby, 1976.
Dockter, Warren. *Churchill and the Islamic World*. London: I. B. Tauris, 2015.
Earhart, Amelia. *Last Flight*. New York: Crown, 1988.
Ellison, Norman. *Flying Matildas: Early Days in Australian Aviation*. Sydney: Angus and Robertson, 1957.
Evans, Colin. *A Question of Evidence*. New York: Barnes & Noble, 2005.
Forden, Lesley. *The Ford Air Tours 1925-31*. New Brighton, MN: Aviation Foundation of America, 2003.
Gunn, John, *The Defeat of Distance: Qantas 1919-1939*. St Lucia, Queensland: University of Queensland Press, 1969.
Gwynn-Jones, Terry. *Heroic Australian Air Stories*. Sydney: Rigby, 1981.
Gwynn-Jones, Terry. *Pioneer Aviator: The Remarkable Life of Lores Bonney*. St Lucia, Queensland: University of Queensland Press, 1988.
Haynes, Jim, and Jillian Dellit. *Great Australian Aviation Stories*. Sydney: ABC Books, 2006.
Jessen, Gene Nora. *The Powder Puff Derby of 1929: The True Story of the First Women's Cross-Country Air Race*. Naperville, IL: Sourcebooks, 2002.
Livingston, John H. *One-Two: The Story of the Fifth National Air Tour as Related by the Winner*. Troy, OH: Waco Aircraft, 1930.
Mackenzie, Roy. *Solo: The Bert Hinkler Story*, Sydney: Ure Smith, 1979.
Mann, Sheila. *The Girls Were Up There Too: Australian Women in Aviation*. Canberra, ACT: Department of Aviation/Australian Government Publishing Service, 1986.
Matowitz, Thomas G, *Cleveland's National Air Races*. Chicago: Arcadia, 2005.
McIver, Stuart B. *Murder in the Tropics: The Florida Chronicles*, vol. 2. Sarasota, FL: Pineapple Press, 2008.
Polk, William R. *The Elusive Peace: The Middle East in the Twentieth Century*. Abingdon, UK: Routledge, 2013.
Spicer, Chrystopher J. *Great Australian World Firsts: The Things We Made, the Things We Did*. Sydney: Allen & Unwin, 2011.
Tagliacozzo, Eric. *The Longest Journey: Southeast Asians and the Pilgrimage to Mecca*. New York: Oxford University Press, 2013.
Thaden, Louise McPhetridge. *High, Wide and Frightened*. Fayetteville: University of Arkansas Press, 2004.
Walker, Mike. *Powder Puff Derby: Petticoat Pilots and Flying Flappers*. Chichester, UK: John Wiley & Sons, 2003.
Walton, Nancy Bird. *My God! It's a Woman*. Sydney: HarperCollins, 2002.
Watts, Cedric. *A Preface to Conrad*, 2nd ed. New York: Routledge, 2014.
Wels, Susan. *Amelia Earhart: The Thrill of It*. Philadelphia, PA: Running Press, 2009.
Winters, Kathleen. *Amelia Earhart: The Turbulent Life of an American Icon*. New York: Palgrave Macmillan, 2010,
Wood, Gillen D'Arcy. *Tambora*. Princeton: Princeton University Press, 2014.

Magazine Articles

Black, Prudence. "Fashion Takes Flight: Amy Johnson, Schiaparelli and Australian Modernism." *Hecate* 35 (2009): 57-76.
"The Blue Comet: The Seashore's Finest Train." *Lancaster Dispatcher* (Ed Mayover, ed.) 42, no. 1, 1-2.
Brandish, C. R. "Captain W. N. Lancaster: An Adventurous Airman." *Table Talk* 26 (April 1928): 11.
Brown, Margery. "Why Women Pilots Have Organized." *Aeronautics*, March 1930, 35.
"Captain W. N. Lancaster's Fine Flight." *Flight*, 19 January 1928, 39-40.
McCarthy, Larry. "The Alexander 'Eaglerock' Saga.," *Vintage Aeroplane*, September 1978, 8-15.
Miller, Ralph N. "Remember the Bullet?" *Popular Aviation*, January 1939, 33, 76-78.
Neville, Leslie E. "The Fleet Biplane." *Aviation*, 18 May 1929, 1686-90.
Patterson, Alicia. "I Want to Be a Transport." *Liberty*, 7 September 1929, 18-24.
Reinhold, Ruth M. "The Old Douglas Airport." *Journal of Arizona History* 15, no. 4 (Winter 1974): 325-48.
Richardson, Eudora Ramsay. "The Ladies of the Lobby. *North American Review* 227, No. 6. (June 1929): 648-655.
"The Robinson Redwing." *Motor Sport*, September 1930, 38-40.
"Tragedy Falls from Sky into Life of Woman Flyer." *The Arrow*, 29 April 1932, 2.
"With an Avro 'Avian' ('Cirrus' Mk.III) in America." *Flight*, 14 March 1929, 215.

Additional Material

Barker, Ralph. "Chubbie Pugh." Transcripts of interviews conducted c. 1968.
Carson, James. M., and Francis P. Malone. *The Lancaster Case: Arguments for the Defense*. Miami: 1934.

Newspapers

Australia

Adelaide Chronicle
The Advertiser [Adelaide]
The Age [Melbourne]
The Argus [Melbourne]
The Barrier Miner [Broken Hill]
Brisbane Courier
Cairns Post
Canberra Times
The Catholic Press
Courier Mail [Brisbane]
Daily Mercury [Mackay]
Daily News [Perth]
Daily Standard [Brisbane]
Evening News [Sydney]
The Examiner [Launceston]
Freeman's Journal [Sydney]
Gippsland Times
Longreach Leader
The Mail [Adelaide]
Maryborough Chronicle
The Mercury [Hobart]
Morning Bulletin [Rockhampton]
The New Call [Perth]
Newcastle Morning Herald and Miner's Advocate
Newcastle Sun
The News [Adelaide]
Northern Standard [Darwin]
The Observer [Adelaide]
Queensland Times
The Queenslander
The Register [Adelaide]
Saturday Night Telegraph [Brisbane]
Sporting Globe [Melbourne]
Sunday Mail [Brisbane]
Sunday Times [Sydney and Perth]
Sydney Morning Herald
The Telegraph [Brisbane]
Townsville Daily Bulletin
Truth [Sydney]
The West Australian
The Worker [Brisbane]

Canada

Border Cities Star [Windsor, Ontario].
Calgary Daily Herald
Montreal Gazette
The Ottawa Citizen

New Zealand

The Press [Canterbury]

Singapore

Straits Times

United Kingdom

Daily Express
Daily Mirror
Yorkshire Post

United States

Brooklyn Daily Eagle
Evening Gazette [Xenia, Ohio]
Florence Times [Florence, Alabama].
Free-Lance Star [Fredericksburg, Virginia].
Lewiston Daily Sun [Lewiston, Maine]
Matawan Journal [New Jersey]
Milwaukee Journal
Milwaukee Sentinel
Nashua Daily Telegraph [Nashua, New Hampshire]
New York Times
News and Courier [Charleston, South Carolina]
Pittsburgh Post-Gazette
Pittsburgh Press
Rochester Evening Journal [New York]
St. Petersburg Times [Florida]
Schenectady Gazette [New York]
Utica Daily Press [New York]
Wichita Eagle

Index

Abbeville airfield, France 28
Aboukir (Abu Qir), Egypt 34
Adams, Mary 157
Aero Club of America 84
Aero Club of New South Wales, Australia 71, 74
Agra, India 45
Air Force Cross 15, 47, 222
Air Transport Auxiliary 230
Airco DH9 15
Airspeed Courier 219
Akyab (now Sittwe), Myanmar 48
Alexander Aircraft Company 136, 137
Alexander Bullet *118*, 136–138, 139, *141*, 142, *143*, 147–148, 150–151, 152
Alexander Eaglerock 136
Allahabad, India 45
American Eagle airplane 125
American Girl 26
Amman, Jordan 35
Anderson, Keith 96
Andrews, Col. Ivo 27
Andros Island, Bahamas 148–149, 150, 152, 153
Anson 22
Atambua, Indonesia 63, 66
Atkinson, Judge H.F. 181, 182, 186, 190, 191, 196, 197, 200, 201, 204, 209
Atlanta, Georgia 128–129
Atlantic 26, 100, 135, 154
Australia 1, 3, 4, 5, 11, 15, 17, 18, 20, 21, 26, 33, 34, 44, 50, 59, 65, 68, 71, 74, 87, 94, 95, 101, 121, 134, 153, 157, 181, 184, 193, 210, 213, 214, 215, 219, 220, 222, 224, 236
Australian Aero Club 1
Australian Flying Corps 16
Australian Light Horse 16
autogyro 222–223
Avro (A.V. Roe and Company) 22, 45, 59, 60
Avro, "Baby" 60
Avro Avian 581 (Hinkler's) 60, 67
Avro Avian 594 1, 5, 20, 22, 25, 26, 46, 49, 52, 53, 91–92, 93, 213, 214–215, 216, 223, 224
Avro 504 22
Azores 25

Bailey, Lady Mary 213
Bailey, Major 33, 34
Bandar Abbas, Iran 41
Bangka Island, Dutch East Indies (now Indonesia) 5, 7, 57–59, 62
Baghdad, Iraq 15, 34, 35, 37
Barbados 93
Barker, Ralph 2
Barnard, Charles 217
Barnes, Florence "Pancho" 98, 105, 110, 112
Basrah, Iraq 35, 38, 39, 40
Bass Strait, Australia 3, 82, 83, 85, 230
Batten, Jean 219, 220
Bayley, Arthur 9
Beech, Walter 101
Bell, Lawrence "Larry" 98, 99
Benghazi, Libya 31, 32
Bennett, Jim 15
Bermuda 90
Bermuda Triangle 147, 148
Beveridge, Beatrice 11
Beveridge, Charles Stanley 9, 10, 11, 14
Beveridge, Eleanor Jean 11, *12*
Beveridge, Emma 11
Beveridge, Ethelwyn Maude 13, 76, 81, 132, 138–139, 150, 153, 155, 157, 206–207
Beveridge, Herbert 11
Beveridge, Jessie 11
Beveridge, Nellie 11
Beveridge, Rev. Sidney Alexander 11
Beveridge, Thomas (Tommy) 13, 14
Beveridge, Rev. Thomas George 11
Black, Victor 11
Blue Comet (train) 94
Bonney, Lores 220
Booth, J.R. 132
Bournemouth, U.K. 19
Bowman, William 141
Brancker, Sir Sefton 20, 23, 24, 27
Brinsmead, Col. H. 80
Brisbane, Queensland 71, 72, 74, 75
British Airways 229, 230
British Empire 33, 74, 91

258 Index

British Empire Airways 231
Broken Hill, New South Wales 13
Brown, Margery 120, 121
Brown, Peggy 173, 175, 192, 193, 202
Bruce, Prime Minister Stanley 27, 68, 74, 75
Bruce, Hon. Mrs. Victor 221–222, 228
Bryant, Millicent Maud 1, 94
Bundaberg, Queensland 60
Bushehr, Iran 40, 41

Cairo, Egypt 34, 35
Calcutta (now Kolkata), India 46–48, 50
Camooweal, Queensland 70
Canada 3, 121, 122, 124, 132–33
Canberra, Australian Capital Territory 77–8, 79, 87
Cape Town, South Africa 91, 213, 215, 217, 223, 224
Carson, James Milton *179*, *180*, 181, 183, 184–206, 209
Casey, Lady Maie 121
SS *Cashmere* 11
Castrol 23
Catania, Italy 30
Cates, J.G. 145
Cave-Brown-Cave, Group Capt. Henry M. 34, 43–44
Cavill, Percy 149–150
Centenary Air Race 219–221
Cessna (airplane) 129
Cessna, Clyde 101
Chabahar, Iran 41
Charleston, South Carolina 144
Charleville, Queensland 69, 71
Chautauqua 156
Churchill, Winston 39
Cincinnati, Ohio 129
Cirrus engine 92, 132
City of Portsmouth 221
Civil Aeronautics Authority (America) 136
Civil Aviation Department (Australia) 75, 79, 80
Clarke, Dr. Beverly 198
Clarke, Haden *162*; character 185, 195–196, 200, 203, 207; death 4, 170–173, 203, 234; exhumation and autopsy 181, 199; funeral 174; romance with Jessie 3, 161–170, 186, 193–194; skull exhibited 197; spirit 206
Clarke, Ida Clyde (nee Gallagher) 160–161, 171, 173, 174, 176, 181, 185, 194, 197, 199, 206
Cleveland, Ohio 96, 97, 98, 101, 103, 113, 114, 117, *119*, 157, 209
Clifford, Sir Hugh 55
Clifford, Lady 55
Clover Field, Santa Monica 99, 100, 105
Cobham, Sir Alan 15, 17, 25, 219
Cochran, Jacqueline 219, 220
Columbus, Ohio 109, 117, 138, 141
Commercial Air Hire Company 228, 229
Commercial Bank 10, 12
Congregational Church 11

Conrad, Joseph 5
Coolgardie, Western Australia 9, 11
Coral Gables, Miami, Florida 158–159, 162
Costes, Dieudonne 26
Cotonou, Benin 226–228
Craighead School, Timaru, New Zealand 13
Cross, Bert 85
Crossland, Charlie 9
Crosson, Marvel 98, 100, 103, 106, 108–109, 119
Croydon Airfield, U.K. 22, 23, 24, 25, 26, 89, 134, 213, 224, 228
Cuba 3, 23, 141–146
Curtiss Airport, Chicago 130, 139
Curtiss, Glenn 130
Curtiss Company of Cuba 145
Curtiss Field, Valley Stream, Long Island 90, 120, 139, 140, *141*
Curtiss Robin (airplane) 160, 161
Curtiss-Wright Corporation 135, 150
Cypress Airways 231

Darwin, Northern Territory 1, 7, 56, 60, 63, 65–69, *67*, 80, 134, 219
The Dawn 26
De Chair, Lady Enid 71, 76
Deederer, Dr. Carlton 171–172, 173, 174–175, 203
De Havilland DH50 15
De Havilland DH60 Gipsy Moth 20, 25, 23, 125, 134, 139, 219
De Havilland DH84 Dragon 222–223
Detroit, Michigan 124, 130, 133
Dillenz, Lilli 25
RRS *Discovery II* 11
Doak, William N. 209–210
Dole Air Race 26
Doolittle, James 97
Doran, Mildred 26
Douglas, Arizona 107–108
Drake, Sir Francis 11
Dudley, Anne Dallas 161
Dum Dum Airport, Calcutta (now Netaji Subhas Chandra Bose International Airport, Kolkata) 47
Dutch East Indies (now Indonesia) 5, 18, 57–59, 62–63

Earhart, Amelia 1, 3, 24, 92, 93, 94, 96, 98, 100, 102, 105, 107, 108, 109, 111, 112, 113, 114, *115*, 117, 119, 120, 121, 123, 135, 137, 139, 155, 161, 226
Edison, Thomas 125
Egypt 32
El Paso, Texas 109, 110, 166, 188, 189, 190
Elder, Ruth 26, 98, *104*, 108, 111, 112, 114, *115*
Elliot, Arthur 15, 25
Empain, Baron 34
Endeavour 26
England 1, 3, 4, 5, 6, 11, 14, 15, 20, 23, 26, 28, 34, 35, 42, 44, 47, 49, 60, 74, 82, 84, 92, 189, 204, 210, 211, 217, 219, 220, 223, 228, 231, 236
HMS *Enterprise* 40, 43, 44

Index

Equator, crossing 7, 57, 236
Essendon Aerodrome, Melbourne 79–80, 83
Euphrates River 38

Fairchild, Sherman Mills 122, *123*
Fairchild Aircraft Company 122
Fairchild Aviation Center and Flying Field (now Republic Airport, Long Island) *123*
Fairchild KR-34 122, *123*, 125
Fairchild White Fleet 122
Fairey (company) 22, 23
Fannie Bay Gaol, Darwin 65–66
Far East 3, 6
Federation Aéronautique Internationale 102, 139
Fleet (airplane) 98–99, 100, 101, *116*, *118*
Fleet, Reuben H. 98
Fleet Aircraft (company) 98, 119
Flying Fish 90
Fokker F-VII 217
Fokker F-VIIb trimotor 87, 89
Follett, Evelyn 1
Folz, Edith 103, *104*, 110, 112, *118*, 136, 137
Ford (company) 123
Ford, Edsel 123, 124
Ford, Henry 123, 125
Ford Airport, Dearborn 125
Ford Tour 1929 122–131, 132; *see also* National Air Tour
Ford trimotor 89, 123, 125, 130
Forrester, James 209, 210
Forsythe, Commissioner Elgin 149, 150
Fort Plain, New York 156
Fort Worth, Texas 111–112, 189
French, Jack 158, 159
Friendship 1, 135
Fysh, Hudson 70

Ganges River 46, 48
Gao, French Sudan (now Mali) 215, 216, 225–226, 234, 235
General Machado Airport, Havana *145*
Giles, Frederick 26
Gillingham, U.K. 10
Gillis, Fay 120
Glencross, Mrs. A. 80, 81
Gloucester House 220
Gower, Commander Pauline 230
Granger, Clema 99
Granger, Jim 99
Grayson, Frances 26
Great Bell of Dhammazedi 50
Gulf of Oman 38

Haizlip, Mary 100, 105, 106, 111, 112, 114, 125, 126, 127, 128, 129
Haldeman, Capt. George 26
Hall, Ethelwyn Maude Lavinia 11
Hall, Jean 11
Hall, Rev. J.K. 11
Halliburton, Erle P. 112

Hamilton, Albert H. 197–199, 200, 203–204
Hanworth Aerodrome, U.K. 221
Haring, J.V. 177
Harper, Jean 153
Harrell, Frances 120, 125, 128, 129, 209
Harris, Sir Arthur "Bomber" 17
Havana, Cuba 141, 142, *143*, 144, *145*–146, 147, 151, 153
Hawaii 26, 88
Hawks, Frank 125, 142
Hawthorne, N. Vernon 173, 174, 175, 183–204
Hayes, Sam 61
Heath, Lady Mary 91–92, 119
Heinrich, Albert S. 131
Heliopolis, Egypt 34
Heston Aerodrome, U.K. 223, 228
Hinaidi *37*, 38
Hinchliffe, Capt. Walter 26
Hinkler, Bert 27, 59–61, 66–67, 68, 69, 71, 74, 75, 79, 85, 152
Hitchcock, Bob 96
Hobart, Tasmania 72, 82–5
Hong Kong 34
Hudson, Earl 172–173, 186, 192, 202, 203
Hunter, Capt. Thomas 11
Huston, Ernest 171–172, 185, 186, 194

Imperial Airways 28, 35, 36, 42
Ince, Joseph 194, 199
India 17, 18, 23, 28, 34, 35, 47, 217
Ingalls, Laura 135, 139, 140, 141, 147, 219
Iraq 39
Ireland 26
Irrawaddy River 49
Italian Air Force 29, 31, 32
Italy 31

Jacksonville, Florida 128, 144, 152
Jakarta, Indonesia 62
Jodhpur, India 45
Johnson (later Mollison), Amy 1, 3, 84, 134, 213, 214, 216, 219, 220, 230
Jolley, Mayor W.A. 72, 74
Jones, Henry M. 200, 201
Junkers trimotor 25

Kalgoorlie, Western Australia 9, 12
Karachi, Pakistan 25, 40, 42, 44
Khan, King Amanullah 42, 43
Kidman, Sir Sidney 20
Kookaburra 95
Kreidner-Reisner Aircraft Company 122
Kuala Lumpur, Malaysia 55
Kunz, Opal 101, 105, 106, 112

Lancaster (bomber) 22
Lancaster, Annie Maud "Kiki" (nee Mervyn-Columb) 17, 19, 27, 60, 71, 76, 80, 82, 87, 90–91, 92, 131, 155, 163, 213, 218–219, 234
Lancaster, Edward 16, 21, 27, 76, 82, 184, 214, 217–218

260 Index

Lancaster, J.K. 79
Lancaster, Maud (Sister Red Rose) 16, 21, 22, 27, 76, 83, 85, 210, 215, 218, 235, 236
Lancaster, William "Bill": arrested 177; background 15–17; diaries 159, 160, 163, 166, 167–168, 175–176, 178, 189–190, 204, 209, 233–236; feelings towards Jessie 163–164, 166–169, 189, 190, 193, 236; first meeting with Jessie 15–17; flight from England to Australia 18–65, *67*; flight from London-Sahara-Cape Town 213–218; flight from New York to Georgetown 92–93, 95; flight within Australia 66–87, *72, 73, 77, 86*; trial 182–208
Lathero, James 175, 177, 179, 206, 209
Latin-American Airways 158, 164, 166, 171, 172, 175, 186, 188, 190
Launceston, Tasmania 82, 83, 85
Lawrence, T.E. 15
Le Bourget Field, Paris 16, 28
Le Brix, Joseph 26
Leete, Bernard 23, 47, 48, 50
Liggett, John 141, 150, 152, 153
Lindbergh, Charles 15, 101, 135, 177
Lindop, Squadron Leader Victor Erskine 34
Livingston, John H. 126, 128, 131
Lockheed Vega 103, 106, 111, 129, 153, 157, 220
London, U.K. 21, 24, 60, 89, 91, 95, 211, 219, 228
London Missionary Society 11
Long Beach, California 139
Long Island, New York 90, 92, 120–121, 133–134, 139, 141
Lord Jim 5
Los Angeles, California 89, 90, 92, 95, 99, 110, 140, 153, 157
Lowenstein, Princess 80
Lympne Aerodrome, U.K. 27, 215, 216, 222
Lyon, Harry 87, 88, 89, 90, 92, 95, 155
Lyons, France 28

Machado, Gen. Gerardo 146
Mackay, Honorable Elsie 26, 80; *see also* Wyndham, Poppy
Mackenzie, Donald 14
Madagascar 11
Maddux, Jack 157
Maddux Airlines 157
Malta 30
Marriott, Sir Hayes 6, 7, 56, 57, 59
Marseilles, France 29
Mascot Aerodrome, Sydney 75
Maungmagon Beach, Myanmar 53–54
McIntosh, Lt. John Cowe 15, 17, 67
McMaugh, Hilda 1, 94
McQueen, Elisabeth Lippincott 98
Mediterranean Sea 3, 18, 29, 30, 31, 35, 215
Melbourne, Victoria, Australia 7, 13, 51, 78–82, 86, 93, 154, 155, 219, 221
Meteorological (Met.) Office flights, U.K. 229

Miami, Florida 92, 114, 144, 145, 146–147, 150, 151, 152, 158, 168–209
Middle East 3, 18
Millar, Ebenezer 51
Miller, Jessie Keith "Chubbie": American citizenship attempts 151, 154, 212; and Bill Lancaster 15–17, 19, 160–170, 174, 182, 191, 192, 213, 216–217, 232–236; breaking records 47, 56, 57, 65, *67*, 68, 74, 81, 131, 134, 135, 139, *141, 142, 143, 145*, 146; childhood 11–13; clothing and fashion *64, 67*, 72, *73*, 74, 78, 79, 83–4, 122, *123*, 125, 219; deportation from America 210–211; divorce from Keith Miller 154–156; family background 10–11; flights and races (American transcontinental record 139–*141*; Centenary Air Race 219–221; Darwin-Hobart 66–85; England-Australia 5–8, 18–65; England-West Africa 224–228; National (Ford Reliability) Air Tour of 1929 122–131; National Air Races, Cleveland *119–120*, 139; Pittsburgh-Havana 141–153, 143, 145; Women's National Air Derby 1929 98–120, *104, 115, 116, 117*); ground crew 47, 55, 68–9, 70, *73*; Haden Clarke 161–174, 182, 188, 190–191, 212; lecturing 81–82, 85, 86, 87, 95, 156; marriage to Keith Miller 13; marriage to John Pugh 229; Ninety-Nines formation 120–121; pilot 72, *77, 73*, 83, *86*, 89, 93–94, 100, 101, *104, 115, 116, 117, 118, 119*, 131, 132–33, 134, 138, *141*–153, *145*, 160, 208, 211, 221, 224–228; pilot's licenses 93–94, 132–133, 221; proposal for women transport pilots 230; snake in the plane 52–53; trial witness 186–188, *187*, 190–191, 212; woman 71, 74, 75, 77, 78, 80, 81, 83, 125, *154, 187*, 208
Miller, Keith 13, 14, 20, 51, 76, 79, 80, 150, 154–156
Miss Doran 26
Miss Long Beach **118**
Mission of the Flowers 16, 21
Moe, Deputy U.S. Marshal 196
Monkey Point, Myanmar 50, 51
Montreal, Canada 126
Mooney, Al 136
Morrell, Sir Stephen 80
Morris, Capt. Jack 23
Mount Tambora, Indonesia 62–63
Mount Vesuvius, Italy 30
Muntok (Mentak), Bangka Island 7–8, 57–59, 61, 62
Mussolini, Benito 30, 43

Naples, Italy 30
Nashville, Tennessee 161, 194
Nassau, Bahamas 150, 151, 182
National Aeronautic Association (America) 139
National Air Races 96, 97, 98, 113, *119, 120*, 139, 157, 209
National Air Tour 1929 122–131, 132; *see also* Ford Tour

Index

New Jersey 92, 93–94, 95, 131, 154
New Zealand 13, 26, 219
New York 89, 92, 93, 94, 95, 122, 131–132, 133–134, 138–139, 140, 142, 153, *154*, 158, 159–160, 161, 164, 210–211
New York state 94, 135, 156, 197
Newcastle, New South Wales 75
Newfoundland 135, 154
Nichols, Ruth 111, 117, 153, 154, 219
Nile River, Egypt 35
Ninety-Nines 3, 4, 120–121, 153, 230
North Africa 31
Noyes, Blanche 109, 110–111, 112, 113, 114, *115*, 117, *118*
Nymagee, New South Wales 153

Oates, John Valentine 11
Oates, Thomas Herbert Beveridge 11
O'Donnell, Gladys 111, 114, *115*, 117, *118*, 139
Ohio 1, 96, 98, 116–121
Omlie, Phoebe 98, 109, 110, 111, 114, *115*, 119, 120
Oran, Algeria 215, 216

Palestine 35
Parer, Lt. Ray 15, 17, 67
Paris, France 26, 28, 154, 228
Paris, Neva *104*, 114, 120
Parks Field and Air College, St Louis 114, *115*, *116*, *117*, *118*
Pasni (now in Pakistan) 42
Patna, India 46
Pecos, Texas 110, 112
Percival, Edgar 220
Percival Gull 220
Perry, Margaret *104*, 109, 112
Persia 40
Perth, Western Australia 9, 10, 13
Philadelphia 23
Phoenix, Arizona (Sky Harbor Airport) 107
Pinxton, Nottingham, U.K. 11
Pisa, Italy 29
Pittsburgh, Pennsylvania 141, 142, *143*, 144, 145, 146, 152, 153
Plymouth, U.K. 34
Point Cook, Melbourne, Australia 75
Portland, Maine 127
Portland, U.K. 11
Post, Wiley 97, 101, 127, 129
Powder Puff Derby 1929 3, 103
Pugh, John Barnard 221–223, 228–229, 230
Pulitzer, Ralph 97
Pulitzer Speed Trophy 97
Pulitzer Trophy Race 97
Putnam, George 3, 90, 93

Qantas 69, 70, 71
Queensland, Australia 70–75
Quimby, Harriet 84

R100/101 airships 42
Raffles Hotel 56

Rangoon (Yangon), Burma (Myanmar) 48, 49, 51, 53
Rasche, Thea 100, 102, *104*, 106, 109, 113, 114, *115*, 119
Red Bank, New Jersey 93–94
Red Rose 1, 5–8, 23, 26, 29, 32, 33, 35, 36, *37*, 40, 42, 44, 45, 46, 47, 49–50, 51, 54, 56, 59, 61, 63, *64*, *67*, 68, *72*, *73*, 76, *78*, 79, 85, 86, 87, 95–96, 155, 215, 230
Reed, Carl *123*
Reggane, Algeria 215, 216
Riga, Latvia 27, 60
Risely, Thomas 9
Robertson, Sir MacPherson 219, 220
Robinson Aircraft Company 223
Robinson Redwing airplane 223–228
Rodgers Airport, Philadelphia 23
Roe, Alliott Vernon 22
Rogers, Will 103
Rome, Italy 30, 31, 60
Rooke, Capt. Dennis 25
Roosevelt Field, Long Island 16, 94, 97, 127, 139, 140
Royal Aero Club 20
Royal Air Force 16, 17, 18, 30, 34, 35, 37, 38, 42, 59, 222, 230
Royal Air Force Far East Flight 34, 40, 43–44, 61
Royal Australian Air Force 75, 80
Royal Naval Air Station Donibristle, HMS *Merlin* 231
Royal Naval Air Station Hinstock, HMS *Godwit* 231
Royal Naval Air Station Stretton, HMS *Blackcap* 231
Royal Naval Volunteer Reserve 231
Royal Navy 11, 30
Russell, Jack 158, 164, 166, 167, 188–189, 193, 194, 203
Russell, Dame Mary, Duchess of Bedford 217
Rutba Wells 35, 36
Ryan monoplane 15
Ryland, Mervyn 15
Ryrie, Sir Granville 20, 27
Ryrie, Lady Mary 27

Sacco-Vanzetti murder case 197, 199
Sahara Desert 2, 214, 216, 225, 232
St. Kilda, Melbourne 13
St Louis, Missouri 114, *115*, 129, 157, 194
Samoa 11
San Bernardino, California 103–105, *104*
San Francisco, California 26
Santa Monica, California 96, 97–98, 99, 109
Savoia Marchetti airplane 132, 133–134
Saw, Charles 10
Schiaparelli, Elsa 24, 122
Seletar, Singapore 61
Serangoon Road racecourse 7
Shell (company) 62, 67, 70, 72, 79, 81, 129
Shelton, Gentry 157, 158, 159–160, 164, 166

262 Index

Shiers, Wally 15
Shwedagon Pagoda 49, 50
Sikorsky seaplane 150
Sing, Maharajah Umaid 45
Singapore 6, 18, 34, 45, 56, 59–62, 71, 230
Singapore Flying Club 6, 7, 56
Sirte, Libya 31
Smith, Sir Charles Kingsford 26, 87, 88, 95, 213, 222
Smith, Elinor 92, 139
Smith, Harwood 141, 142, 144
Smith, Sir Keith Macpherson 15, 17, 62, 67, 76
Smith, Sir Ross MacPherson 15, 17, 62, 67
Sopwith Aviation Company 60
Southern Cross 87, 95
Southern Cross constellation 9
Southern Cross Minor 213–214, 236
Southern Cross, Western Australia 9, *10*, 11–12, 106
Spinks, Pasha Maj.-Gen. Sir Charlton 33
Spirit of St. Louis 15
Stack, Capt. Thomas Neville 23, 47
Stanford, Edward 21
Starr, Margaret 14, 15
Stearman, Lloyd 101
Stinson, Eddie, and aircraft company 133
Stinson, Katherine 133
Stinson Detroiter 26
Stinson Junior 132, 133
Stork, Charles T. & aircraft company 132–133
Straits Times 6
Sumatra, Indonesia 7
Supermarine Southhampton II flying boats 34, 40, 43–44, 61; *see also* RAF Far East Flight
Surabaya, Indonesia 62
Sydney, New South Wales 1, 71, 74, 75–*77*, *78*, 86, 95, 230

Taiping, Malaysia 55
Tallman, Dr. 198–199
Tancrel, Mark 158, 164, 190, 193, 194, 196–197, 199, 203
Tasmania 3, 82–85
Taj Mahal 45
Taylor, Florence 86
Test Pilot (Dir. Victor Fleming. Star. Clark Gable and Spencer Tracy) 97
Thaden, Louise 98, 100, 103, *104*, 106, 108, 109, 111, 113, 114, *115*, 117, *118*, 119, 209, 219
Thompson, Freda 1, 24
Tiber River 30
Tigris River 37, 38
Timaru, New Zealand 13
Timor, Indonesia 63
Timor Sea 3, 7, 18, 56, 63–64
Toomey, Mick 9
Toowoomba, Queensland 71, *72*, *73*
Toronto, Canada 125
Trans-Continental & Western Airlines (TWA) 157

Travel Air (airplane) 100, 101, 106, 110, 113, *118*, 142
Tripoli, Libya 30, 31
Trout, Bobbi 106, 109, 112, 114, 116, 118–119, 209
Tulsa, Oklahoma 112

Ulm, Charles 87, 95
United States 3, 4, 26, 87, 88, 91, 92, 94, 121, 124, 126, 135, 141, 147, 151, 153, 173, 181, 184, 209–210, 228
United States Department of Commerce 102
United States Department of Labor 4, 181, 209–210
Upton, Frank 194
Ur Junction, Iraq (near Babylonian Ur) 38–39

Vaughan, Guy 135
Vickers Victoria 34, 35
Vickers Vimy 15
Victor Aircraft Corporation 131
Victoria, Australia 219
Victoria Point (Kawthaung), Myanmar 53, 54
Voelter, Karl 150, 151
Von Mach, Mary 105, 107, 114, *115*
Vulcan (bomber) 22

Waco (airplane) 93, 128, 131, 131
Wakefield, Sir Charles (and Wakefield company) 23, 45, 57, 72, 74, 76
Walker, Mike 14, 101
Walker, Vera Dawn *104*, 105, 108, 112, *115*
Walton, Nancy Bird 3, 121, 230–231
Wangaratta, Victoria, Australia 11
Washington, D.C. 139, 152, 153, 209, 210
Watts, Mayor Douglas 66, 68, 69
Weekly Times 13
Western Australia 9, *10*, 12, 13, 95, 155
Wichita, Kansas 110, 112, 113, 114, 129, 139, 140
Wignall, Mayor J. 84
Williams, Errett 136, 137
Women: aviation fashion 114, 122–123, 125, 219; Aviation Pioneer Hall of Fame 4; as pilots 91, 94, 102, 107, 112, 120–121, 125, 133, 139, 142, 215, 219–220, 230; as public figures 207; suffrage 161
Women's National Air Derby 1929 3, 97, *104*, *118*, 136; *see also* Powder Puff Derby
World War I 11, 15, 16, 22, 23, 131
World War II 4, 11, 17, 230, 231
Wright engine 100, 101, 122, 131, 135, 137
Wyndham, Poppy 26; *see also* Mackay, Hon. Elsie
Wynyard, Tasmania 85

Xenia, Ohio 1, 116

Yeargin, O.C. 172, 186
Yilgarn, Western Australia 9
Yuma, Arizona 105, 106, 112

Ziza (Al Jiza) 35

www.ingramcontent.com/pod-product-compliance
Ingram Content Group UK Ltd.
Pitfield, Milton Keynes, MK11 3LW, UK
UKHW041932140426
5217IPUK00014B/436